Prealgebra
Thinking Like a Mathematician

A Work-Text for
Pre Algebra

Second Edition

Jim Symons
De Anza Community College
Cupertino, California

The McGraw-Hill Companies, Inc.
Primis Custom Publishing

New York St. Louis San Francisco Auckland Bogotá
Caracas Lisbon London Madrid Mexico Milan Montreal
New Delhi Paris San Juan Singapore Sydney Tokyo Toronto

McGraw-Hill Higher Education
A Division of The McGraw-Hill Companies

Prealgebra
Thinking Like A Mathematician
A Work-Text for Pre Algebra

Copyright © 1995 by McGraw-Hill, Inc. All rights reserved. Printed in the United States of America. Except as permitted under the United States Copyright Act of 1976, no part of this publication may be reproduced or distributed in any form or by any means, or stored in a data base retrieval system, without prior written permission of the publisher.

McGraw-Hill's Primis Custom Publishing consists of products that are produced from camera-ready copy. Peer review, class testing, and accuracy are primarily the responsibility of the author(s).

10 11 12 13 14 15 QSR QSR 0 9 8 7 6 5 4 3 2

ISBN 0-07-062167-5

Editor: Todd Bull
Cover Design: Mark Anderson
Printer/Binder: Quebecor World

Contents

Introduction v

Chapter 1 **Integers**
- 1.1 Operations on Whole Numbers 1
- 1.2 Operations on Integers 11
- 1.3 Review of Operations on Integers 21
- 1.4 Review of Operations Integers and Introduction of Translations 27
- 1.5 Chapter Review 35

Chapter 2 **Equations and Word Problems**
- 2.1 Introduction to Equations 43
- 2.2 Solving Equations 51
- 2.3 Solving Equations and Word Problems 59
- 2.4 Equations with Variables on Both Sides of the Equal Sign 63
- 2.5 Solving Equations and Word Problems Using the Distributive Property 69
- 2.6 Chapter Review 77

Chapter 3 **Applications of Equations**
- 3.1 Solving Word Problems 83
- 3.2 More Word Problems 91
- 3.3 Word Problems Involving Sum of Two Numbers 95
- 3.4 Problem Set 101
- 3.5 Chapter Review 103

Chapter 4 **Fractions**
- 4.1 Introduction to Fractions 109
- 4.2 Operations on Common Fractions I 119
- 4.3 Operations on Common Fractions II 129
- 4.4 Operations on Mixed Numbers 135
- 4.5 Translations Involving Fractions 145
- 4.6 Evaluating Algebraic Expressions Using Fractions 153
- 4.7 Addition or Subtraction with Large Denominators 159
- 4.8 Word Problems Having Fractions with Large Denominators 163
- 4.9 Chapter Review 167

Chapter 5   Equations with Fractions

     5.1   Solving Equations with Fractions   177

     5.2   Using the Distributive Property on Equations with Fractions   185

     5.3   Ratios and Proportions   191

     5.4   Word Problems Using Proportions   201

     5.5   Chapter Review   207

Chapter 6   Percent

     6.1   Decimals and Use of a Calculator   215

     6.2   Introduction to Percent   225

     6.3   Practice with Percent   233

     6.4   More Practice with Percent   239

     6.5   Chapter Review   243

Chapter 7   Graphs

     7.1   Introduction to Graphs   249

     7.2   Making Graphs   259

     7.3   The Function Notation   267

     7.4   Chapter Review   275

Chapter 8   Geometry

     8.1   Using Formulas from Geometry: Rectangles and Triangles   281

     8.2   Using Formulas from Geometry: Circles   287

     8.3   Using Formulas from Geometry: Pythagorean Theorem   291

     8.4   Using Geometric Formulas on Irregular Figures   299

     8.5   Dealing with Dimensions   307

     8.6   Chapter Review   311

Chapter 9   Final Review Problem Sets

     9.1   Problem Set   317

     9.2   Problem Set   321

     9.3   Problem Set   325

Weights and Measures   **329**
Useful Formulas   **330**
Answers to Selected Problems   **331**
Index   **379**

Introduction

In this course, you will learn to think like a mathematician.

You are about to embark on an adventure in mathematics. You will **not** be asked to perform the tedious task of multiplying large numbers or adding long columns of numbers. These are tasks that can be done more easily and more accurately by a calculator. Instead, you will do the thinking and organizing of a mathematician that will enable you to let the calculator do the tedious arithmetic. We will call this "thinking like a mathematician."

Thinking like a mathematician means you must:

1. **Organize information**
2. **Look for patterns**
3. **Make decisions**
4. **Master skills**

Let us begin with some problems that will illustrate the process.

◆ **Example 1**   You are offered a job with two different options for your salary. Option 1 would pay you $20,000 per year to start and an increase of $2,000 every year after the first year. Option 2 would pay you $10,000 per six months to start and an increase of $500 every six months after the first six months. Which option pays more money?

The non-trained mathematician might rush to some answer, but this is a problem deserving careful consideration and certainly some **organization** of information in

Organization of Information

Option 1: Salary		Option 2: Salary		
1st year	$20,000	1st 6 months	$10,000	
		2nd 6 months	$10,500	
		Total		$20,500
2nd year	$22,000	3rd 6 months	$11,000	
		4th 6 months	$11,500	
		Total		$22,500
3rd year	$24,000	5th 6 months	$12,000	
		6th 6 month	$12,500	
		Total		$24,500

Look for a Pattern

The pattern here is one of addition. By properly organizing this information, one can more easily see the **pattern** that each year Option 2 with its six month increment will always yield an additional $500 more than the Option 1.

Make a Decision

Based on the information of the problem, one should choose Option 2

◆ **Example 2** You have a rectangular shaped lot which measures 45 meters by 30 meters. You would like to fence the entire area and fence three equal rectangular lots as inexpensively as possible so that you could rent these out as horse stalls. Find the total cost of the least expensive plan if the fence costs $23 per meter to build.

Organization of Information

For this problem, drawing a sketch might help you see the nature of the problem. In fact, there are two different possible drawings for this problem.

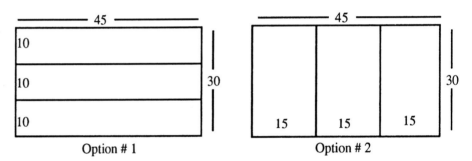

Look for a Pattern

The pattern here is one of addition. To find the length of the outside fence, add the four sides: 45+30+45+30 = 150 meters. Then if the inside fences run parallel to the longer side as in Option 1, you would need two inside fences of 45 meters each. 45+45 = 90 meters of inside fence. If the inside fences run parallel to the shorter sides as in Option 2, you would need two inside fences of 30 meters each. 30+30 = 60 meters of inside fence.

Make a Decision

The **pattern** clearly indicates that the smaller amount of fence would be needed if the inside fences run parallel to the shorter sides. The cost would be calculated by multiplying the total length of fence by the cost per meter.

Cost per meter times total number of meters of fencing is the total cost.

$$\$23\,(\,150\,+\,60)\,=\,\boxed{\$4830}$$

So, clearly you would decide that Option 2 is the least expensive.

Rules to Live By

1. Believe in yourself! Learning a new skill is hard work and at times can be very frustrating, but you can do it. If at first, you don't succeed, try, try again. Don't give up!

2. Attend every class and be on time.

3. Work every problem in the text.

4. Check your answers in the Appendix and rework every problem you have missed.

5. If help is available through your school or through any other resource, take advantage of it. Don't try to be machoman or superwoman. Smart people take advantage of the help available.

6. After a test, rework every problem that you have missed. Mathematics is a cumulative subject; that is, one uses all or many of the skills previously learned in the mastering of a new concept. You cannot afford to miss any topics or skills along the way.

7. Work neatly! Copy problems very carefully. Work down and line up the terms as you bring them down. Do not bunch up your work. Leave plenty of room so that you can read what you have done.

8. Do not skip steps. Do not work several steps on top of each other. Until you are an expert, it is important that you take great care with writing down every step. This is important not only for your good work habits, but also for receiving credit for the correct work that you have done.

9. On word problems, think about your answers. Ask yourself: Does the answer make sense? For example, what if a problem asks for the amount of change received from a sales transactions? You know that the amount of change can never be greater than the amount of money you handed the cashier. [Unless the cashier failed math and is giving you the incorrect amount.] Also, try estimating the answer whenever possible. Ask yourself: Is your answer close to your estimate?

10. Work collaboratively. The word collaborative is the combination of the prefix "co" meaning together and the root word "labor." In other words, collaborate means to work together. Throughout history, famous scientists and artists have worked collaboratively to achieve great results in their professions. While it is important that you first attempt to learn the materials on your own, it will be helpful if you check your assignments and review lessons with your fellow students.

11. Have fun and good luck! And remember: the harder you work—the luckier you get!

CHAPTER 1
Integers

In this chapter, you will learn to:

1. *Perform the operations of addition, subtraction, multiplication and division on integers.*
2. *Use the correct order of operations to simplify an expression.*
3. *Evaluate an algebraic expression given values of the variables.*
4. *Translate English expressions into mathematical expressions.*
5. *Solve word problems using integers.*

1.1 Operations on Whole Numbers

In this course you will be expected to **think like a mathematician.** You will **organize information, look for patterns, make decisions,** and **develop skills.** Throughout the pages of this book you will be given instruction, examples and problems to give you practice in this mathematical process.

This first section begins with whole numbers and develops an organized approach to simplifying mathematical expressions. While this chapter deals with the group of numbers called "integers," this first section will involve only the subgroup of integers called "whole numbers." The concept of integers will be explained in section 1.2.

What are Whole Numbers?

The set of numbers beginning with zero, and all numbers that can be derived by successively adding one to the previous number is called the set of **whole numbers** (0, 0+1, 1+1, 2+1, 3+1, 4+1, ...). We can partially list the set as:

$$\{ 0, 1, 2, 3, 4, 5, 6, 7, 8, 9, 10, 11, 12...\}$$

Pictured on a number line, the whole numbers are graphed below:

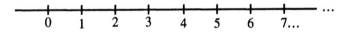

The three dot notation ... means that the pattern is to be continued and is read as "and so on." Notice that the number 0 is the smallest number in the set, but there is no largest whole number. We use the whole numbers to count "whole" objects, with the number zero being the "absence of a count."

What is a Mathematical Expression?

A mathematical expression is a phrase using numbers. The phrase may use the basic arithmetic operations of addition, subtraction, multiplication, division, powers or roots. If the expression uses only numerals we call it a **numerical expression**. If the expression uses a combination of numerals and letters to represent numbers we call it an **algebraic expression**. The letters used in an algebraic expression will be called **variables**.

Examples of numerical expressions:

$$5 + 3 \qquad 8 \div 2 \qquad (4)(3) - 6 \qquad \sqrt{9} \qquad 5^2$$

Examples of algebraic expressions:

$$5a \qquad 3x - 2y \qquad 8v \div 4b \qquad 8x + 63s + 4s$$

When an algebraic expression contains a numeral next to a variable with no indicated operation the understood operation is multiplication. For example 5x means "five times x." The numeral in this expression is called the **numerical coefficient** of the expression.

A **constant** is a is a symbol that represents a **known quantity**. For example:

$$5 \qquad 8 \qquad -3 \qquad \frac{2}{3} \qquad 3^2$$

are all constants.

A **variable** is a symbol that represents an unknown quantity. We usually represent variables with letters such as a, b, c, x, y and z.

Order of Operations

We shall begin with a few **expressions** which should be simplified. Sometimes more than one operation is contained in an expression. When several operations appear in the same expression, one must follow the "**order of operations**" rule. In general, one works from left to right and does the operations in the following order:

1. **Parenthesis** (perform the operation within the grouping symbols such as parenthesis, brackets, square root symbols, or fraction bars)
2. **Exponents**
3. **Multiplication** or **division** (whichever comes first working left to right)
4. **Addition** or **subtraction** (whichever comes first working left to right)

An **exponent** tells how many times the **base** number is used as a **factor**. In the expression 6^2 the base number is 6 and the exponent is 2. So $6^2 = (6)(6) = 36$. Other examples:

$$4^3 = (4)(4)(4) = 64 \qquad\qquad 2^5 = (2)(2)(2)(2)(2) = 32$$

You can see that an exponent indicates repeated multiplication. Also note that when no operation symbol is given multiplication is the understood operation. For example, (3)(4) means "3 times 4."

Helpful Hint 1: *To remember the order of operations, try the mnemonic device:* **"Please Elect My Dear Aunt Sally."** *The first letter of each word stands for an operation.* **P =** *parenthesis or grouping.* **E =** *exponents.* **M =** *multiplication.* **D =** *division.* **A =** *addition.* **S =** *subtraction. If you need to multiply and divide in the same problem, do the operation that appears first working left to right. Do the same for addition and subtraction.*

◆**Example 1**   Simplify the expression: $5 + 2(9 - 4)^2$

In this expression we see addition, multiplication, grouping and exponents and therefore we will simplify using order of operations as seen below. (Remember: **Please Elect My Dear Aunt Sally**)

$$5 + 2(9-4)^2$$

$= 5 + 2(5)^2$	Subtraction in the parenthesis	Please
$= 5 + 2(25)$	Exponents	Elect
$= 5 + 50$	Multiplication	My dear
$= \boxed{55}$	Addition	Aunt Sally

◆**Example 2** Simplify the expression: $\quad 5^2 - 3^2\left(\dfrac{24}{6} - \dfrac{14}{7}\right)$

(Remember: **Please Elect My Dear Aunt Sally**)

$$5^2 - 3^2\left(\dfrac{24}{6} - \dfrac{14}{7}\right)$$

$= 5^2 - 3^2(4-2)$	Division in the Parenthesis
$= 5^2 - 3^2(2)$	Subtraction in the Parenthesis
$= 25 - 9(2)$	Exponents
$= 25 - 18$	Multiplication
$= \boxed{7}$	Subtraction

Helpful Hint 2: WORK DOWN! *Notice in the above two examples that the problem was worked down - not sideways. In the second step, the 5 was "brought down" directly below the 5 in the first step. The negative sign (–) was brought down directly below the negative sign (–) in the first step, and so on. "Working down" and lining up the terms helps you copy correctly the terms that are merely being "brought down" and also helps you check the result of the operations.*

Evaluating an Algebraic Expression

An expression containing variables is called an **algebraic expression.** If we know the value of the variables, we can evaluate the expression by replacing the variables with given values, and then follow the "order of operations." As you do these problems, begin by recopying the problem with all of the indicated operations in place but using an empty parenthesis every time a variable appears. Then fill in the empty parenthesis with the given values of the variables.

◆**Example 3**   Evaluate the expression: $a^2 - bx$ if $a = 6$ $b = 4$ $x = 3$

$a^2 \;-\; b\,x$	Copy carefully
$= (\;)^2 - (\;)(\;)$	Write an empty parenthesis for each variable
$= (6)^2 - (4)(3)$	Fill in the empty parenthesis
$= 36 \;-\; (4)(3)$	Exponents
$= 36 \;-\; 12$	Multiplication
$= \boxed{24}$	Subtraction

Helpful Hint 3: The second line above showing empty parenthesis should not appear on your paper after you have completed the problem. It is shown above to demonstrate how you should first replace each variable with a parenthesis and then fill in that parenthesis with the value given.

◆**Example 4**   Evaluate the expression: $ax + b[bx - a]$ if $a = 3$ $b = 5$ $x = 2$

$ax + b[bx - a]$	
$= (3)(2) + (5)[(5)(2) - (3)]$	Empty parenthesis, fill in. (use bracket instead of parenthesis because it looks less confusing)
$= (3)(2) + 5[10 - 3]$	Multiplication in Parenthesis
$= (3)(2) + 5[7]$	Subtraction in Parenthesis
$= 6 + 35$	Multiplication
$= \boxed{41}$	Addition

Helpful Hint 4: Write down every step and do not crowd your work. Leave plenty of space between terms so that you can see what you are doing. Do not work several steps on top of each other on the same line. It's too easy to make mistakes that way.

Word Problems

Word problems are one of the best devices for developing your skill to **think like a mathematician.** In order to solve word problems, you must **organize the information** that is given. You will need to decide what you know, what you don't know (the unknown), what is given but not important. As you saw in the Introduction to this book, you may need to categorize the information given.

The next step is **look for patterns.** You will recognize many patterns, because of your everyday experiences and your common sense. You know that words you use every day have mathematical equivalents, like "increased by" means added to or +, decreased by means to subtract, times means to multiply, to separate something into equal parts means to divide.

You will have to **make decisions** about what information should be used and what operations will be used on this information and finally use the **skills** you have developed to find the answer required in the problem.

◆ **Example 5** While on vacation, Alice rented a Moped at a basic cost of $12 plus $6 per hour of use. She rented the Moped for a total of 5 hours and paid for it with a $50 travelers check. How much change did she get?

1. **Organize** information:

    Alice must pay $12 in addition to $6 for each of the 5 hours of use.

2. Look for **patterns**:

    The cost is determined by adding the basic cost ($12) to the cost of using it for 5 hours.

3. Make **decisions**:

    Cost is $12 + 5($6)

4. Use **skills** to simplify the expression:

    Cost is $12 + $30 or $42.

5. **Answer the question:** Since the question asked for the amount of change from a $50 travelers check, we must decide to perform a subtraction to obtain the required answer.

    $50 − $42 = $8

The question is then answered with a complete sentence:

> Alice received $8 in change.

◆**Example 6** When Fidel was 23 years old, he paid $4 for his math pencil and with this marvelous writing instrument scored 88, 95, 98 and 91 on the next four tests. What was his average test score?

1. **Organize** information.

    Fidel's age and cost of his pencil are of no interest in this problem. The only information we need are his test scores of 88, 95, 98, and 91

2. Look for **patterns**.

    The pattern here is to add the scores together and then divide by the number of tests which is 4.

3. Make **decisions** which in this case means write an expression.

    $(88 + 95 + 98 + 91) \div 4$

4. Use **skills** to simplify this expression.

    $(88 + 95 + 98 + 91) \div 4$

    $= 372 \div 4$

    $= 93$

5. **Answer the question** with a sentence.

    > Fidel had a test average of 93.

1.2 Operations on Integers

In section 1.1, you did operations on **whole numbers.**

Set of **whole numbers** = { 0, 1, 2, 3, 4, 5, 6, 7, 8, ...}

In this section, you will do operations on **integers.**

What are Integers?

Set of Integers = { ..., –3, –2, –1, 0, 1, 2, 3, ...}

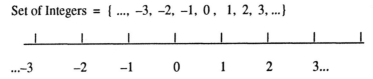

The set of **integers** is an extension of the whole numbers, as it contains all of the whole numbers in addition to all of the opposites of the whole numbers. The number "–2" is called the **opposite** of 2, or the **negative** of the number 2.

– 5 is the opposite or negative of 5.

–287 is the opposite or negative of 287.

8 is the opposite of –8.

What is the Absolute Value of a Number?

In order to operate on integers we need to understand the concept of **absolute value.**

The **absolute value** of a number is the <u>distance</u> **from** zero **to** that number on the number line.

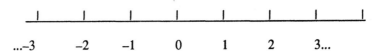

For example, the integer "–3" is three units from zero. So, the absolute value of –3 is 3 and is written as $|-3| = 3$. The number 3 is also three units from zero so $|3| = 3$.

Other examples:

The integer –5 is five units from zero so	$|-5| = 5$
The integer 6 is six units from zero so	$|6| = 6$
The integer –10 is ten units from zero so	$|-10| = 10$

Addition of Integers

To perform the operation of addition on two integers, first consider the **signs** of the integers. There are two different rules depending on whether the signs are alike or different.

Rule 1. Rule for addition of integers with like signs

If the signs are **alike**, <u>add</u> the absolute values and use the common sign.

Rule 2. Rule for addition of integers with different signs

If the signs are **different**, <u>subtract</u> the smaller absolute value from the larger absolute value and use the sign on the number with larger absolute value.

◆**Example 1** Addition of integers with **like** signs.

$$(-5) + (-3) = ?$$

Because the signs are alike, add the absolute values and use the negative sign because it is common to both numbers.

$$(-5) + (-3)$$
$$= -[|-5| + |-3|] \quad \text{Use negative sign, add the absolute values}$$
$$= -[5 + 3]$$
$$\boxed{-8}$$

Other examples:

$$(-6) + (-7) \quad = -13$$
$$(1) + (8) \quad = 9$$
$$(-23) + (-4) \quad = -27$$

◆**Example 2** Addition of integers with **different** signs.

$$(7) + (-12) = ?$$

Because the signs are different, take the difference of the absolute values and use the sign of the number with the larger absolute value.

$$(7) + (-12)$$
$$= -[|-12| - |7|] \quad \text{Use negative sign, because sign on 12 is negative.}$$
$$= -[12 - 7]$$
$$\boxed{-5}$$

Other examples:

$$(-3) + 7 \quad = 4$$
$$(16) + (-21) \quad = -5$$
$$15 + (-7) \quad = 8$$

Subtraction of Integers.

The operation of subtraction is an inverse operation.

Rule 3. Subtraction is the addition of the first number to the opposite of the second

◆**Example 3** Subtraction: $5 - 8 = ?$

$5 - 8$ means 5 added to the opposite of 8 or symbolically: $5 + (-8)$

Refer back to the rule of addition of integers with different signs. This rule says to subtract the smaller absolute value from the larger absolute value and use the sign of the integer with the larger absolute value.

Section 1.2: Operations on Integers

$$5 + (-8)$$
$$= -(|-8| - |5|)$$

Subtract the smaller absolute value **from** the larger absolute value and use the sign of the integer with the larger absolute value.

$$= - (|-8| - |5|)$$

$$\boxed{-3}$$

◆**Example 4** Subtraction: $-8 - 2 = ?$

$$-8 - 2 \text{ means } (-8) + (-2)$$

Refer back to the rules for Addition when signs are alike. This rule says to add the absolute values and use the common sign.

$$(-8) + (-2)$$
$$= -(|-8| + |-2|)$$
$$= -[8 + 2]$$
$$\boxed{-10}$$

◆**Example 5** $-9 - (-3)$ means $-9 + (3)$

Subtraction means "add the first number to the opposite of the second number."

$$= -(|-9| - |3|)$$

Signs are different. Take the difference of the absolute values, use the sign of the integer with the larger absolute value

$$= -(9 - 3)$$
$$\boxed{-6}$$

◆**Example 6** 5 – hot means 5 + the opposite of hot

5 – hot = (5 + cold)

Multiplication of Integers

To perform the operation of multiplication on **two** integers we again must first consider the **signs** on the integers. There are two different rules depending on whether the signs are alike or different.

Rule 4. Rule for multiplication of two integers when the signs are alike.

If the signs are **alike**, multiply the absolute values and the **product is positive.**

Rule 5. Rule for multiplication of two integers when the signs are different.

If the signs are **different**, multiply the absolute values and the **product is negative.**

14 Chapter 1: Integers

◆ **Example 7** Multiplication of integers with **like** signs

$$(-3)(-5) =$$

$$= +(|-3|) \cdot (|-5|)$$ Multiply the absolute values of the two numbers and use the **positive** sign

$$= +(3)(5)$$

$$\boxed{15}$$

If the number is positive, no + sign is needed

◆ **Example 8** Multiplication of integers with **different** signs

$$(-2)(7)$$

$$= -(|-2|) \cdot (|7|)$$ Multiply the absolute values of the two integers and use the **negative** sign.

$$= -(2)(7)$$

$$\boxed{-14}$$

◆ **Example 9** $(-4)(6) = -24$ Signs are **different**, multiply absolute values, the product is **negative**.

◆ **Example 10** $(-8)(-5) = 40$. Signs are the **alike**, multiply absolute values, the product is **positive**.

Division of Integers

To perform the operation of division on **two** integers, we must again look at the signs of the integers. There are two rules depending on whether the signs are alike or different.

Rule 6. Rule for division of <u>two</u> integers if the signs are alike.

If the signs are **alike**, divide the absolute values and the quotient is **positive.**

Rule 7. Rule for division of <u>two</u> integers if the signs are different.

If the signs are **different**, divide the absolute values and the quotient is **negative.**

◆ **Example 11** Division of integers with **like** signs

$$(-24) \div (-6)$$

$$= +(|-24| \div |-6|)$$ Use the **positive** sign and divide the absolute value of the two integers.

$$= +[24 \div 6]$$

$$= \boxed{4}$$ If answer is positive, no "+" sign is needed.

◆ **Example 12** Division of integers with **different** signs

$$(-35) \div (5)$$

Section 1.2: Operations on Integers   15

$= -(|-35| \div |5|)$ Use the **negative** sign and divide the absolute value of the two numbers..

$= \boxed{-7}$

◆**Example 13** $36 \div (-9) = \boxed{-4}$ Signs are **different** so quotient is **negative**.

◆**Example 14** $(-42) \div (-7) = \boxed{6}$ Signs are **alike** so quotient is **positive**.

Order of Operations on Integers

Review: Remember the mnemonic device for order of operations: Please Elect My Dear Aunt Sally. Also, remember that you must first complete all the operations within the parenthesis (or any kind of grouping) before doing the operations outside of the parenthesis. When simplifying the expression within the parenthesis, use the correct order of operations: Exponents, Multiplication, Division, Addition and Subtraction.

◆**Example 15** $-8 + 3\left[(-5)^2 - (7)(5)\right]$

$= -8 + 3[25 - (7)(5)]$ Parenthesis: Exponents

$= -8 + 3[25 - 35]$ Multiplication

$= -8 + 3[-10]$ Subtraction

$= -8 + (-30)$ Multiplication

$\boxed{-38}$ Addition

Evaluating an Algebraic Expression Using the Set of Integers

Review: Copy the expression using empty parenthesis in place of each variable, and then fill in the parenthesis with the given value of the variable.

◆**Example 16**   Evaluate $b - n[ax - nb]$ if $a = -3$, $b = 4$, $n = -2$, and $x = -4$

$b - n[ax - nb]$

$= 4 - (-2)[(-3)(-4) - (-2)(4)]$ Fill in the parenthesis

$= 4 - (-2)[12 - (-8)]$ Parenthesis: Multiplication of integers with like signs and different signs

$= 4 - (-2)[12 + 8]$ Parenthesis: Subtraction

$= 4 - (-2)[20]$ Parenthesis: Addition

$= 4 - (-40)$ Multiplication

$= 4 + 40$ Subtraction

$\boxed{44}$ Addition

◆**Example 17**   Using the same values as in Example 16, evaluate $x + \dfrac{ab}{n} - x^2$

$$x + \dfrac{ab}{n} - x^2$$

$= (-4) + \dfrac{(-3)(4)}{(-2)} - (-4)^2$ Fill in the parenthesis

$= -4 + \dfrac{-12}{-2} - (-4)^2$ Grouping: Multiplication (different signs)

$= -4 + \dfrac{-12}{-2} - (\mathbf{16})$ Exponents

$= -4 + \mathbf{6} - 16$ Division (like signs)

$= \mathbf{2} - 16$ Addition (different signs)

$= 2 + (-16)$ Subtraction means add the opposite

$\boxed{-14}$ Addition (different signs)

1.3 Review of Operations on Integers

In this section, we will continue thinking like a mathematician by organizing information, looking for patterns, making decisions and developing skills. Our **organizational** tasks include working down, writing clearly, not bunching up and boxing in the answer. We will use the **patterns** established in the rules for operating on integers. We will make **decisions** as we decide which operation to do next based on the order of operations established in section 1.1 and we will use the **skills** of arithmetic to obtain correct answers.

So far the two most important concepts that we need to master are the order of operations and the rules for operation on integers. For the order of operations we need to remember: "Please Elect My Dear Aunt Sally." (Parenthesis, exponents, multiply or divide and finally add or subtract.) The rules for the operations on integers are listed in the table below:

Table 1.1 Rules for Operations on Two Integers

If the operation is...	and the signs are...	do this operation...	and the answer will have this sign...
Addition	Alike	Add the absolute values	Common sign
	Unlike	Subtract smaller absolute value from larger absolute value	Sign of number with larger absolute value
Subtraction—Use definition: subtraction means add the opposite. Use the rules for addition.			
Multiply	Alike	Multiply absolute values	Positive
	Unlike	Multiply absolute values	Negative
Division	Alike	Divide absolute values	Positive
	Unlike	Divide absolute values	Negative

The following example emphasizes the organization used in evaluating an algebraic expression:

◆ **Example 1** Evaluate the expression $x + a[b - aw]$ if $a = 4$, $b = -3$, $x = -2$, $w = 1$

$$x + a [b - aw]$$
$$= (\ ) + (\ )[(\ ) - (\ )(\ )] \qquad \text{Write empty parenthesis}$$
$$= (-2) + (4)[(-3) - (4)(1)] \qquad \text{Fill in with given values}$$
$$= -2 + 4[(-3) - 4] \qquad \text{Multiplication within the } \textbf{Parenthesis}$$
$$= -2 + 4[-3 + (-4)] \qquad \text{Subtraction means add the opposite}$$
$$= -2 + 4[-7] \qquad \text{Addition within the } \textbf{Parenthesis}$$
$$= -2 + (-28) \qquad \text{Multiplication}$$
$$\boxed{-30} \qquad \text{Addition}$$

1.4 Review of Operations Integers and Introduction of Translations

In this lesson we will continue to practice our skill in operating on integers and we will begin to learn the important skill of translating words into symbols. The skill of correctly translating from English into algebraic symbols will be helpful in solving word problems.

Words Into Symbols

There are many words used in the English language to describe the four basic operations of addition, subtraction, multiplication and division.

Words for Addition.

How many words can you think of that can be translated to the **operation of addition**? Many words that you use in everyday conversation translate into the operation of addition. Words like "**sum of**," "**more than**," "**added to**," "**greater than**," "**older than**" translate into the mathematical symbol of "**+**."

Examples:

The **sum** of 2 and 6	means 2 + 6 or 6 + 2
5 **more than** 4	means 4 + 5 or 5 + 4
3 **added to** 2	means 2 + 3 or 3 + 2
1 **greater than** 6	means 6 + 1 or 1 + 6
2 years **older than** 8	means 8 + 2 or 2 + 8

Note: The operation of addition is said to be **commutative**. Commutative means that the order of performing the operation of addition does not change the result of the addition. As seen above, 2 + 6 is the same as 6 + 2 and 4 + 5 is the same as 5 + 4, etc.

Words for Subtraction.

Words like "**difference**," "**younger than**," "**diminished by**," "**less than**," "**decreased by**," "**subtract....from**" are some of the words used for subtraction or the mathematical symbol of "**−**."

The **difference of** 9 and 2	means 9 − 2
3 years **younger than** 7	means 7 − 3
8 **diminished by** 5	means 8 − 5
4 **less than** 9	means 9 − 4
6 **decreased by** 4	means 6 − 4
Subtract 8 **from** 2	means 2 − 8
Subtract 2 **from** 8	means 8 − 2

Note: Subtraction is **not commutative**, which means there is only one way to translate these expressions. As you can see 9 − 2 would result in a different answer than 2 − 9.

Words for Multiplication

Words like **"product," "times," "twice,"** and **"multiply"** are some of the words that translate into the mathematical symbol "•" or no symbol at all such as 5x which means 5 • x or 5 times x.

Examples:

The **product** of 3 and 5	means 3 • 5 or 5 • 3
6 **times** 7	means 6 • 7 or 7 • 6
twice x	means 2x
y **multiplied by** 8	means 8y

Note: The operation of multiplication is **commutative**, which means that the order of performing the operation of multiplication does not change the final product. As in the examples above, we see that 3 • 5 is the same as 5 • 3 and 6 • 7 is the same as 7 • 6 In an algebraic expression like 2x, it is not wrong to write x2 but for clarity we will always write the numerical coefficient first. A **numerical coefficient** is the constant that is multiplied by the variable.

Words for Division

"Quotient," "divided by" and **"ratio of"** are some of the words that translate into the mathematical symbol "÷" or the standard fraction bar as seen below.

Examples:

The **quotient** of 12 and 3	means $12 \div 3$ or $\frac{12}{3}$
24 **divided** by 8	means $24 \div 8$ or $\frac{24}{8}$
the **ratio of** 5 to 3	means $5 \div 3$ or $\frac{5}{3}$

Note: The operation of division is **not commutative**. $12 \div 3$ is **not** the same as $3 \div 12$.

Words for Equal

Verbs such as **"is," "will be"** and **"was"** often translate into the mathematical symbol, "=."

Examples:

The sum of 5 and 4 **is** 9	means $5 + 4 = 9$
12 decreased by 4 **will be** 8	means $12 - 4 = 8$
The product of 5 and 3 **was** 15	Means $(5)(3) = 15$

Use of Variables for Unknown Numbers

You may choose any letter or any symbol to represent the unknown quantity so long as it is not confusing. For example the letter "o" is a bad choice because it may be confused with the number "0."

Section 1.4: Operations on Integers and Translations

Examples of Translating Words into Mathematical Symbols

◆ Example 1 Translate and simplify:

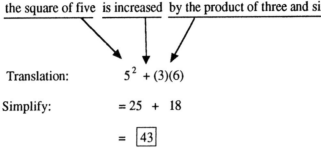

Find the result when the square of five is increased by the product of three and six.

Translation: $5^2 + (3)(6)$

Simplify: $= 25 + 18$

$= \boxed{43}$

◆ Example 2 Translate and simplify: Find the result when the quotient of ninety and three is diminished by three times the cube of two.

Translation: $\dfrac{90}{3} - 3(2)^3$

Simplify: $= 30 - 3(8)$

$= 30 - 24$

$= \boxed{6}$

Examples of Translating Words into Mathematical Symbols Using Variables

In mathematics, we often must deal with quantities whose values we do not know. By choosing a variable to represent the unknown quantity, we can translate an English expression containing an unknown number into mathematical symbols.

◆ Example 3 Translate:

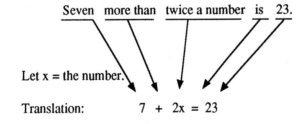

Seven more than twice a number is 23.

Let x = the number.

Translation: $7 + 2x = 23$

◆ Example 4 Translate: Nine less than four times a number is 15.

Let y = the number

Translation: $4y - 9 = 15$

◆ Example 5 Translate: When six times a number is decreased by seven the result will be 41.

Let a = the number

Translation: $6a - 7 = 41$

◆**Example 6** Translate: If three times a number is increased by nine the result is 45.

Let n = the number,

Translation: $3n + 9 = 45$

◆**Example 7** Translate: Eight less than the quotient of five times a number and four is 17.

Let w = the number,

Translation: $\dfrac{5w}{4} - 8 = 17$

◆**Example 8** Translate: If a number is divided by eight the result is nine more than twice the number.

Let n = the number,

Translation: $\dfrac{n}{8} = 9 + 2n$

◆**Example 9** Translate: The sum of four times a number and nine is six more than the number.

Let x = the number

Translation: $4x + 9 = 6 + x$

1.5 Chapter Review

The numbers in the brackets indicate the sections where the topics were discussed.

Order of Operations [1.1]

To simplify an expression, you must always follows the correct order of operations:

1. **Parenthesis**: Simplify the expression within the grouping symbol such as the parenthesis ( ), brackets [ ] or a fraction bar.
2. **Exponents**: Evaluate any numbers with exponents.
3. **Multiplication and division**: Perform these operations working whichever operation comes first moving left to right.
4. **Addition and subtraction**: Perform these operations working whichever operation comes first moving left to right.

The saying: "Please Elect My Dear Aunt Sally" may help you remember the proper order.

◆**Example 1** Simplify $2(6-3)+3^2$

$$
\begin{aligned}
2(6-3)+3^2 &= 2(3)+3^2 && \text{Subtraction within \textbf{parenthesis} first} \\
&= 2(3)+9 && \text{Exponents} \\
&= 6+9 && \text{Multiplication} \\
&= \boxed{15} && \text{Addition}
\end{aligned}
$$

Evaluation of an Algebraic Expression [1.1]

When evaluating an algebraic expression, remember to copy the problem carefully and **write empty parenthesis** directly below the variable(s). Remember to **work down** and not to skip steps. After you have substituted the given value(s), then simplify using the correct order of operations.

◆**Example 2** Evaluate the expression $b+x[b-aw]$ if $a=4$, $b=7$, $x=2$, $w=1$

$$
\begin{aligned}
b+x[b-aw] && \text{Copy the problem} \\
=(\ )+(\ )[(\ )-(\ )(\ )] && \text{Write empty parenthesis} \\
=(7)+(2)[(7)-(4)(1)] && \text{Fill in with given values} \\
=7+2[7-4] && \text{Multiplication within the \textbf{Parenthesis}} \\
=7+2[3] && \text{Subtraction within the \textbf{Parenthesis}} \\
=7+6 && \text{Multiplication} \\
=\boxed{13} && \text{Addition}
\end{aligned}
$$

Operations on Integers [1.2]

Set of Integers = { .... –3, –2, –1, 0, 1, 2, 3, ... }

The **absolute value** of a number is the **distance from zero** to that number on the number line.

The symbol $|a|$ means the absolute value of a. For example,

$$|3| = 3 \qquad |-5| = 5 \qquad |-238| = 238$$

Table 1.2 Rules for Operations on Two Integers

If the operation is...	and the signs are...	do this operation...	and the answer will have this sign...
Addition	Alike	Add the absolute values	Common sign
	Unlike	Subtract smaller absolute value from larger absolute value	Sign of number with larger absolute value
Subtraction—Use definition: subtraction means add the opposite. Use the rules for addition.			
Multiply	Alike	Multiply absolute values	Positive
	Unlike	Multiply absolute values	Negative
Division	Alike	Divide absolute values	Positive
	Unlike	Divide absolute values	Negative

◆ **Example 3** Addition: When signs are **alike**, add the absolute values and use the common **sign**.

$$(-8) + (-3) = -11$$

◆ **Example 4** Addition: When signs are **different, subtract** the smaller absolute value from the larger absolute value and use the **sign** of the number with the larger absolute value.

$$\begin{aligned} -9 + 6 &= -[|-9| - |6|] \\ &= -[9 - 6] \\ &= \boxed{-3} \end{aligned}$$

◆ **Example 5** Subtraction: Subtraction is the inverse operation of addition. This means we first write subtraction as the addition of the opposite, then use the rules of addition given above.

$$\begin{aligned} & 9 - 13 \\ &= 9 + (-13) \\ &= \boxed{-4} \end{aligned}$$

◆**Example 6**   Subtraction means add the opposite.

$$-12-(-4)$$
$$= -12 + 4$$
$$= \boxed{-8}$$

◆**Example 7**   Multiplication: When signs are **alike**, multiply the absolute value of the two numbers and use the positive sign.

$$(-9)(-3) = 27$$

◆**Example 8**   Multiplication when signs are **different**, multiply the absolute value of the two numbers and use the negative sign.

$$(-9)(4) = -36$$

◆**Example 9**   Division: When the signs are **alike**, divide the absolute values of the two numbers and use the **positive** sign.

$$(-36) \div (-9) = 4$$

◆**Example 10**   Division: When the signs are **different** divide the absolute values of the two numbers and use the **negative** sign.

$$(24) \div (-8) = -3$$

A System for Solving Word Problems [1.1]

Use the following five step process:

1. **Organize** the information given. Disregard unnecessary information.
2. Look for **patterns** in the information.
3. Make **decisions** on using the information to write an expression that reflects the relationship of the numbers.
4. Simplify the expression using mathematical **skills.**
5. Write a **sentence** that answers the question.

◆ Example 11 Jethro is three years older than his sister Billie Sue. For her birthday, he bought her three compact discs at $12 each and a pretty blue cap for $7. If he paid for this purchase with a $50 travelers check left over from his 12 day vacation, how much change did he get back?

1. **Organize** information:

   $12 = Cost of each of 3 compact discs

   $7 = Cost of the cap

   $50 = Amount given to the clerk

2. **Pattern:** Amount given clerk minus the cost of the CDs and cap = change

3. **Decision:** $50 - [$3(12) + 7] = change

4. Use **skills** to simplify:

   $50 - [$36 + 7] = change

   $50 - $43 = change

   $7 = change

5. **Sentence**   Jethro received $7 in change.

Note: There were several bits of extraneous information in this problem. Be sure to only use information that relates to the question asked.

Translating Words into Symbols: [1.4]

To translate from words into symbols, learn and use the words commonly associated with the operations of addition, subtraction, multiplication and division. Choose a variable to represent the unknown number.

◆ Example 12 Translate:

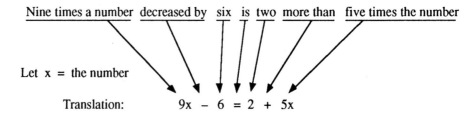

Let x = the number

Translation: $9x - 6 = 2 + 5x$

◆ Example 13 Translate:

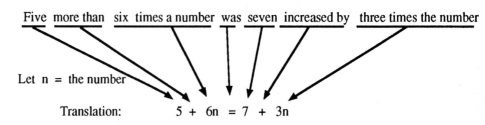

Let n = the number

Translation: $5 + 6n = 7 + 3n$

CHAPTER 2
Equations and Word Problems

In this chapter you, will learn to:

1. Solve linear equations in one variable.
2. Solve number type word problems using linear equations.

2.1 Introduction to Equations

One of the most powerful skills we develop in this course is the skill of solving equations. A few definitions are needed to help us understand the process.

Basic Definitions

As we discussed in Section 1.1, a **constant** is a symbol that represents a known quantity. All integers and fractions are examples of constants. There are other types of numbers that are constants but we will not be using them in this course. Examples of constants are

integers like $-25, \; -3, \; 1 \;$ and $\; 5$

and fractions such as $-1\frac{5}{6}, \; -\frac{3}{5}, \; \frac{2}{3}, \;$ and $\; \frac{5}{4}$.

A **variable** is a symbol that represents an unknown quantity. As you have seen in the last chapter, a variable can have different values from problem to problem. We usually use letters to indicate variables, such as a, b, c ... x, y, and z.

A **numerical coefficient** is a constant that is multiplied by a variable. In essence, the coefficient of a variable tells "how many" of that variable that you have. The coefficient **4** in the term **4x** means that you have **4** of the x's. Examples of coefficients are:

5 is the coefficient of x^2 in the term $5x^2$

23 is the coefficient of y in the term **23y**

$\frac{3}{2}$ is the coefficient of w in the term $\frac{3}{2}w$ which could also be written $\frac{3w}{2}$

$\frac{1}{2}$ is the coefficient of x in the term $\frac{1}{2}x$, which could also be written $\frac{x}{2}$

Note: When the numerical coefficient is "1", the 1 is usually **not** written. For example 1x means one x so we just write "x" instead of 1x. Likewise, "– 1" is usually **not** written as a numerical coefficient. We just write – x instead of – 1x.

A **term** is a constant, a variable or a constant times one or more variables. Examples of single terms are:

$3, \quad x, \quad 3x, \quad 5xy, \quad -16y^2 \quad$ and $\quad -\frac{3}{2}x$

An expression like 5x + 2y has two terms. An expression like 4xy – 2a + 8 has three terms. Notice that the definition of a term involves the operation of multiplication. When terms are separated with addition or subtraction then there is more than one term.

What are "Like Terms"?

When we start to solve equations in Section 2.2, we will be called upon to **combine like terms**, whenever possible. **Like terms** are terms that have the same variable raised to the same powers. Like terms may differ only in their coefficients. Terms that contain only constants are also like terms. Examples of like and unlike terms are given below:

Like terms	**Unlike terms**
5x and 6x	x and x^2
$4y^2$ and $-5y^2$ and $2y^2$	16 and 8x
-16 and $5\frac{1}{2}$	3x and 2y
3xy and 9xy	$3x^2y$ and $9xy^2$

Types of Numbers

In Chapter one we learned how to operate on the set of **integers**.

Integers = { ...., -3, -2, -1, 0, 1, 2, 3, ...}

In this chapter we will introduce the properties of the numbers used in algebra, the **real numbers**. We can think of a real number as any number that can be associated with a point on the number line. As we can see on the number line below, real numbers include integers, fractions, decimals, and roots. Even the number π, used in geometry, is a real number. (approximate value of π is 3.14159, so we see it associated with a point between 3 and 4)

```
-3 ↑  -2   -1   0↑   1↑   2    3↑   4
   -2.6           1/4  √2       π
```

Properties of Real Numbers

We will now state the basic properties of addition and multiplication of real numbers. (Properties for subtraction and division are given as needed but these are defined as inverse operations) These properties may seem rather trivial or self-evident, but they form the basis for developing the concepts of algebra. The following properties are true for all real numbers, represented here by **a**, **b**, and **c**.

1. Commutative property

Rule: For addition, **a + b = b + a**   For multiplication, **ab = ba**

The **commutative property** assures us that we can **add** numbers in any order or **multiply** numbers in any order and the answer will be the same.

For example, $5 + 3 = 3 + 5$ and $(8)(2) = (2)(8)$

Note: The commutative property is **NOT** true for subtraction or division.

For example, $5 - 3 \neq 3 - 5$ and $8 \div 2 \neq 2 \div 8$

The symbol, $\neq$, means is <u>not</u> equal to.

2. Associative property

Rule: For addition, $a + (b + c) = (a + b) + c$ For multiplication, $a(bc) = (ab)c$

The **associative property** is important, because we can only operate on **two** numbers at a time and this property assures us that we will obtain the same end result no matter how we **pair** the numbers for addition or multiplication.

For example, given the problem: $3 + 4 + 5$, we can **add** only two numbers at time.

By **associating** the first two we get	By **associating** the second two we get
$(3+4) = 5$	$3+(4+5)$
$= 7 + 5$	$= 3 + 9$
$= 12$	$= 12$

The above example illustrates that the answer is the same no matter how you **associate** the numbers in addition. The examples below illustrate that the same is true for multiplication.

For example, given the problem: $(2)(3)(4)$, we can **multiply** only two numbers at a time.

One method would be $(2 \cdot 3) \cdot 4$.   Another method would be $2 \cdot (3 \cdot 4)$

$\qquad\qquad\qquad\quad = 6 \cdot 4 \qquad\qquad\qquad\qquad\qquad = 2 \cdot 12$

$\qquad\qquad\qquad\quad = 24 \qquad\qquad\qquad\qquad\qquad\;\; = 24$

The associative property is often used in multiplying terms like $3(4x) = (3 \cdot 4)x = 12x$.

3. Distributive property

Rule: $a(b + c) = ab + ac$ and $a(b - c) = ab - ac$

The **distributive property** enables us to multiply a number by the sum (or difference) of two numbers. The number **a** is multiplied by **b**, the number **a** is multiplied by **c** and then their products are added (or subtracted).

For example, $4(x + 3) = 4x + 4(3) \qquad$ and $\qquad 5(x - 2) = 5x - (5)(2)$

$\qquad\qquad\qquad\quad\;\; = 4x + 12 \qquad\qquad\qquad\qquad\qquad\;\; = 5x - 10$

4. Identity Elements

Rule: $a + 0 = 0 + a$

The **identity** element for addition is <u>zero</u>. This means that zero added to

any number equals the same identical number.

Examples: $5 + 0 = 5 \qquad -8 + 0 = -8 \qquad 0 + (-4) = -4 \qquad x + 0 = x$

Rule: $a(1) = (1)a = a$

The identity element for multiplication is <u>one</u>. This means that any number multiplied by one produces the same identical number.

Examples: $8 \cdot (1) = 8 \qquad -16 \cdot (1) = -16 \qquad (1) \cdot 3 = 3 \qquad x(1) = x$

5. Inverse Properties

Rule: For addition, $a + (-a) = 0$

Because a number minus the same number produces the identity element zero, **subtraction** is said to be the **inverse operation** of addition.

$$a - a = 0$$

Rule: For multiplication, $a \cdot \dfrac{1}{a} = \dfrac{a}{a} = 1, a \neq 0$

Rule for multiplication and division: Because a non-zero number divided by itself produces the identity element one, **division** is said to be the **inverse operation** of multiplication.

$$a \div a = 1, \quad a \neq 0 \quad \text{or} \quad \frac{a}{a} = 1, \quad a \neq 0$$

Addition and Subtraction

$6 + (-6) = 0$

$-5 + 5 = 0$

$8 - 8 = 0$

Multiplication and division

$3 \cdot \dfrac{1}{3} = 1$

$-2 \cdot \left(-\dfrac{1}{2}\right) = 1$

$9 \div 9 = 1$

The following chart summarizes the inverse operations:

Operation	Inverse operation
addition	subtraction
subtraction	addition
multiplication	division
division	multiplication

Combining Like Terms.

To combine like terms involving variables, we must perform the indicated operation of addition or subtraction on the coefficients and multiply the result by the variable.

For example:

$$3x + 5x = (3 + 5)x$$

$$\boxed{= 8x}$$

$$7x^2 - 3x^2 = (7 - 3)x^2$$

$$\boxed{= 4x^2}$$

$\boxed{\text{Note}}$: The operation of adding or subtracting the coefficients of like variables is an example of a use of the **distributive property**.

Section 2.1: Introduction to Equations

◆ Example 1 Combine like terms and simplify: $5z + 3 + 2z + 6$

$5z + 3 + 2z + 6 = 5z + 2z + 3 + 6$ Commutative +

$ = \boxed{7z + 9}$ Combine like terms $5z + 2z$ and $3 + 6$

◆ Example 2 Combine like terms and simplify: $4a - 7 + 3a + 4$

$4a - 7 + 3a + 4 = 4a + 3a + (-7) + 4$ Commutative +

$ = \boxed{7a - 3}$ Combine like terms

◆ Example 3 Combine like terms and simplify: $6 + (-4x) + 2 + x$

$6 + (-4x) + 2 + x = 6 + 2 + (-4x) + x$ Commutative +

$ = \boxed{8 - 3x}$ Combine like terms

◆ Example 4 Combine like terms and simplify: $8 + 2(3x + 5)$

$8 + 2(3x + 5) = 8 + 6x + 10$ Distributive

$ = 6x + 8 + 10$ Commutative +

$ = \boxed{6x + 18}$ Combine like terms

◆ Example 5 Combine like terms and simplify: $a - 5(2a - 3)$

$a - 5(2a - 3) = a - 10a + 15$ Distributive

$ = \boxed{-9a + 15}$ Combine like terms

◆ Example 6 Combine like terms and simplify: $6(2y - 3) - 5(3y - 2)$

$6(2y - 3) - 5(3y - 2) = 12y - 18 - 15y + 10$ Distributive

$ = 12y - 15y - 18 + 10$ Commutative +

$ = \boxed{-3y - 8}$ Combine like terms

Name: Tuyen Tran

2.1 Problem Set 49

Name the property of real numbers used in each of the following:

1) 3 + 5 + 6 = 8 + 6 associative properties of addition.	2) 5(3 + x) = 15 + 5x Distributive properties of Multiplication.	3) 7 + 0 = 7 Inverse properties
4) 9 + (–9) = 0 Inverse properties	5) 7·5·3 = 7·15	6) 4(2x + 3) = 8x + 12
7) $\frac{7x}{7}$ = 1x	8) 1x = x	9) 3·4·5 = 4·3·5

Use the indicated property to write an equivalent statement for each of the following:

10) By commutative + 7 + 3 = 10	11) By inverse + 4 + (–4) = 0	12) By associative • 3·7x = 21x
13) By distributive property 6(x + 4) = 6x + 24	14) By identity • 6(1) = 6	15) By inverse • 3·(1/3) = 1
16) By associatiive + 2 + (4 + 7) = (2 + 4) + 7	17) By distributive property 3(2x – 5) = 6x – 15	18) By commutative • (4·6)·7 = (6·4)·7

Use the distributive property and simplify each of the following:

19) 5(2a +4) = 10a + 20	20) –3(5w –4) = –15w + 12	21) –4(2 +3v) = –8 – 12v
22) 3 + 4(2x +5) 3 + 12x + 20	23) –9 + 2(4y –7) –9 + 8y – 14	24) 3 –4(3a + 2) – 5 3 – 12a + 8 – 5

2.1 Problem Set

Combine like terms on each of the following:

25) $2x + 3 + 7x + 8$
$9x + 11$

26) $5a + 6 + a$
$6a + 6$

27) $3y - 7 - 5y$
$-2y - 7$

28) $4 - 2w - 3w - 5$
$-5w - 1$

29) $-8 + 3k - 2 - k$
$2k - 10$

30) $5a + 2x + 2a - x$
$7a + 1x$

31) $8 + 3(2x - 6) + 7$
$8 + 6x - 18 + 7$
$-10 + 6x + 7$
$6x - 3$

32) $-4a + 3(4a - 3) - 5$
$-4a + 12a - 9 - 5$
$8a - 14$

33) $-6x + 4(2a - 3x)$
$-6x + 8a - 12x$
$-18x + 8a$

34) $-2x - 3x - 4x - 5$
$-9x - 5$

35) $-3 + 4(6x - 3) - 8x$
$-3 + 24x - 12 - 8x$
$16x - 15$

36) $3a + 5z - 6a + z$
$-3a + 6z$

37) $y + 4(2y - 8) + 6$
$y + 8y - 32 + 6$
$9y - 26$

38) $-x - 3(2 - 5x)$
$-x - 6 + 15x$
$14x - 6$

39) $-12 + 3(2a - 5)$
$-12 + 6a - 15$
$6a - 27$

40) $7(x + 5) + 2(x - 6)$
$7x + 35 + 2x - 12$
$9x + 23$

41) $3(2a - 5) - 6(3a - 2)$
$6a - 15 - 18a + 12$
$-12a - 3$

42) $-6(2x + 3a) + 5(4a + 3x)$
$-12x - 18a + 20a + 15x$
$3x + 2a$

2.2 Solving Equations

One of the most useful skills of a mathematician is the skill of solving equations. In this lesson we will learn an organized approach to solving equations.

What is an Equation?

An equation is a statement that two algebraic expressions are equal. Our task will be to solve the equation.

Example: $x + 3 = 8$

left side equals sign right side

Since $x + 3 = 8$ is true when $x = 5$, we call 5 the solution of the equation. To solve an equation means to find all its solutions.

Example: $12 - y = 9$

$12 - y = 9$ is true **only** when $y = 3$ so 3 is the solution of the equation.

Solving Equations by Inspection and by Using Addition and Multiplication Facts

One of the **skills** most often used by mathematicians is that of solving an equation. Some equations can be solved by inspection and recall of simple addition or multiplication facts.

For example, the equation $x + 5 = 8$ can be solved by recalling that $3 + 5 = 8$, so we can say that $x = 3$.

The equation $4x = 20$ can be solved by recalling that $4(5) = 20$, so we say that $x = 5$.

However, when the equation is more complicated like $9x - 7 = 65$ a quick glance may not result in a correct answer. Therefore, we must develop some rules for solving more complicated equations.

Solving Equations by Using the Properties of Equality and the Inverse Operations.

1. The Organization:

To find the solution(s) to an equation, a mathematician would:

1. **organize** the information in a downward step-by-step format;
2. see and use **patterns** of

   * **isolating the variable** on one side of the equation by

   * **producing zero and one** (the "identity elements")

   * using the **properties of equality** for addition, subtraction, multiplication and division and

   * **combining like terms** whenever possible.

2. The Addition Property of Equality

If the same number is added to each side of the equation, the results are still equal.

Rule: If $a = b$, then $a + c = b + c$

Example: $x - 4 = 2$

$x - 4 + 4 = 2 + 4$		Add **4** to both sides.
$x + 0 = 6$		Produce **zero** to isolate the variable
$\boxed{x = 6}$		The solution

We can check the solution by substituting 6 for x in the original equation.

$x - 4 = 2$	Original equation
$(6) - 4 = 2$	Replace x with 6
$\boxed{2 = 2}$	True

3. The Subtraction Property of Equality

If the same number is subtracted from each side of an equation, the results are still equal.

Rule: If $a = b$, then $a - C = b - C$

Example: $x + 4 = 7$

$x + 4 - 4 = 7 - 4$	Subtract **4** from each side.
$x + 0 = 3$	Produce **zero** to isolate x
$x = 3$	Solution
Check: $x + 4 = 7$	Original equation
$(3) + 4 = 7$	Substitute **3** for x
$\boxed{7 = 7}$	True

4. The Multiplication Property of Equality

If each side of an equation is multiplied by the same number, the results are still equal.

Rule: If a = b, then a • c = b • c

Example: $\frac{x}{2} = 6$

$\left(\frac{x}{2}\right)2 = 6(2)$ Multiply each side by **2**

(1) x = 12   Produce **one** x

$\boxed{x = 12}$ The solution

Check: $\frac{x}{2} = 6$ Original equation

$\frac{12}{2} = 6$ Substitute **12** for x

$\boxed{6 = 6}$ True

5. The Division Property of Equality

If each side of an equation is divided by the same number (provided the number is not zero), the results are still equal.

Rule: If a = b then $\frac{a}{c} = \frac{b}{c}$, provided that C ≠ 0

Example: 2x = 12

$\frac{2x}{2} = \frac{12}{2}$ Divide each side by **2**

(1) x = 6   Produce **one** x

$\boxed{x = 6}$ The solution

Check: 2x = 12   Original equation

2(6) = 12   Substitute **6** for x

$\boxed{12 = 12}$ True

6. Combining the Properties of Equality and the Use of Inverse Operations to Undo Equations

An equation is solved by **isolating** the variable on **one side** of the equation.

It does not matter on which side of the equation we isolate the variable. However, for ease of explanation, we will always isolate the variable on the left side of the equation.

Steps to solve an equation by isolating the variable on the left side of the equation are:

1. **Recopy** the problem with **space available** just to the **left of the equal sign** to be able to undo the equation in an orderly manner.
2. **Eliminate** the **constant** term from the left side by filling the space with the inverse operation to produce **zero.**
3. Make the **coefficient** of the variable term **one** by using an inverse operation.

◆**Example 1** Solve the equation: $3x + 4 = 28$

Solution: $3x + 4 = 28$ Recopy with **space available** to the left of the equal sign

$3x + 4 - 4 = 28 - 4$ **Fill space** by subtracting **4** from each side

$3x + 0 = 24$ Produce 0 (eliminate the constant term from the left side)

$3x = 24$ Identity +

$\dfrac{3x}{3} = \dfrac{24}{3}$ Divide each side by **3** (to produce 1x)

$\boxed{x = 8}$ Solution

Check: $3x + 4 = 28$ Original equation

$3(8) + 4 = 28$ Substitute **8** for x

$24 + 4 = 28$ Multiply before you add

$\boxed{28 = 28}$ True

◆**Example 2**   Solve the equation: $9x - 7 = 56$

Solution:

$9x - 7$	$= 56$	**Recopy** with **space** available to left of equal sign
$9x - 7 + 7$	$= 56 + 7$	Add **7** to each side so that the variable term is isolated
$9x + 0$	$= 63$	Work down, produce **0**
$9x$	$= 63$	Identity +
$\dfrac{9x}{9}$	$= \dfrac{63}{9}$	Divide each side by **9**
$\boxed{x = 7}$		Work down, produce **1** x

Check: The solution can now be "checked" by substituting 7 for x..

$9(7) - 7 = 56$

$63 - 7 = 56$

$\boxed{56 = 56}$ The solution is correct.

◆**Example 3**   Solve the equation $6x + 48 = 30$

Solution:

$6x + 48$	$= 30$	Recopy with space to left of equal sign
$6x + 48 - 48$	$= 30 - 48$	Subtract **48** from each side so that the x term is isolated
$6x + 0$	$= -18$	Work down, produce **0**
$6x$	$= -18$	Work down, identity +
$\dfrac{6x}{6}$	$= \dfrac{-18}{6}$	Divide each side by **6**
$\boxed{x = -3}$		Work down, produce **1**x

Check: The solution can now be checked by substituting –3 for x.

$6x + 48 = 30$	Original equation
$6(-3) + 48 = 30$	Substitute **–3** for x
$-18 + 48 = 30$	Multiply
$\boxed{30 = 30}$	The solution is correct.

◆**Example 4** Solve the equation $\frac{x}{5} + 3 = 7$

Solution: $\frac{x}{5} + 3 = 7$ **Recopy with space** to left of equal sign

$\frac{x}{5} + 3 - 3 = 7 - 3$ **Subtract 3** from each side

$\frac{x}{5} + 0 = 4$ Produce **0**, **combine like terms**

$\frac{x}{5} \cdot 5 = 4 \cdot 5$ **Multiply** each side by **5**

$\boxed{x = 20}$ Work down, produce 1x

Check: The solution can now be checked by substituting 20 for x in the original equation. The check is left to the reader.

Name: _____   2.2 Problem Set   57

Solve and check each of the following:

1) $2a + 5 = 11$	2) $7b - 4 = 3$	3) $-3c + 18 = 3$
$2a + 5 - 5 = 11 - 5$	$7b - 4 + 4 = 3 + 4$	$-3c + 18 - 18 = 3 - 18$
$\frac{2a}{2} = \frac{6}{2}$	$7b = 7$	$-3c = -15$
$a = 3$	$\frac{7b}{7} = \frac{7}{7}$	$\frac{-3c}{-3} = \frac{-15}{-3}$
	$B = 1$	$c = 5$
Check:	Check:	Check:
$2(3) + 5 = 11$	$7(1) - 4 = 3$	$-3(5) + 18 = 3$
$6 + 5 = 11$	$7 - 4 = 3$	$-15 + 18 = 3$
4) $-4x - 3 = 5$	5) $-8y + 12 = 20$	6) $2x - 8 = -22$
$-4x - 3 + 3 = 5 + 3$	$-8y + 12 - 12 = 20 - 12$	$2x - 8 + 8 = -22 + 8$
$-4x = 8$	$-8y = 8$	$2x = -14$
$\frac{-4x}{-4} = \frac{8}{-4}$	$\frac{-8y}{-8} = \frac{8}{-8}$	$\frac{2x}{2} = \frac{-14}{2}$
$x = -2$	$y = -1$	$x = -7$
Check:	Check:	Check:
$-4(-2) - 3 = 5$	$-8(-1) + 12 = 20$	$2(-7) - 8 = -22$
$8 - 3 = 5$	$8 + 12 = 20$	$-14 - 8 = -22$
7) $-m + 5 = 18$	8) $\frac{x}{4} + 5 = 8$	9) $\frac{n}{-5} + 6 = 10$
$-m + 5 - 5 = 18 - 5$	$\frac{x}{4} + 5 - 5 = 8 - 5$	$\frac{n}{-5} + 6 - 6 = 10 - 6$
$-m = 13$		$\frac{n}{-5} = 4$
$-m \cdot -1 = 13 \cdot -1$	$4 \cdot \frac{x}{4} = 3 \cdot 4$	$\frac{n}{-5} = 4 \cdot -5$
$m = -13$	$x = 12$	$n = -20$
Check:	Check:	Check:
$-(-13) + 5 = 18$	$\frac{12}{4} + 5 = 8$	$\frac{-20}{-5} + 6 = 10$
$18 = 18$	$3 + 5 = 8$	$4 + 6 = 10$

2.2 Problem Set

---

10) $7w - 9 = -23$

$7w - 9 + 9 = -23 + 9$

$7w = -14$

$\dfrac{7w}{7} = \dfrac{-14}{7}$

$w = -2$

Check:
$7(-2) - 9 = -23$
$-14 - 9 = -23$

---

11) $-5y + 7 = 32$

$-5y + 7 - 7 = 32 - 7$

$-5y = 25$

$\dfrac{-5y}{-5} = \dfrac{25}{-5}$

$y = -5$

Check:
$-5(-5) + 7 = 32$
$25 + 7 = 32$

---

12) $-3r + 40 = -35$

$-3r + 40 - 40 = -35 - 40$

$-3r = -75$

$\dfrac{-3r}{-3} = \dfrac{-75}{-3}$

$r = 25$

Check:
$-3(25) + 40 = -35$
$-75 + 40 = -35$

---

13) $\dfrac{x}{6} - 4 = 1$

$\dfrac{x}{6} - 4 + 4 = 1 + 4$

$\dfrac{x}{6} = 5$

$\dfrac{x}{6} = 5 \cdot 6$

$x = 30$

Check:
$\dfrac{30}{6} - 4 = 1$
$5 - 4 = 1$

---

14) $-2k - 18 = -38$

$-3k - 18 + 18 = -38 + 18$

$-3k = -20$

$\dfrac{-3k}{-3} = \dfrac{-20}{-3}$

$k = 17$

Check:
$-2(17) - 18 = -38$
$-34 - 18 = -38$

---

15) $7 - 2x = 19$

$7 - 7 - 2x = 19 - 7$

$-2x = 12$

$\dfrac{-2x}{-2} = \dfrac{12}{-2}$

$x = -6$

Check:
$7 - 2(-6) = 19$
$7 + 12 = 19$

---

16) $8 - (-3x) = 59$

$8 - 8 - (-3x) = 59 - 8$

$-(-3x) = 51$

$\dfrac{3x}{3} = \dfrac{51}{3}$

$x = 17$

Check:
$8 - [-3(17)] = 59$
$8 - (-51) = 59$
$8 + 51 = 59$

---

17) $4 - 6x = -44$

$4 - 4 - 6x = -44 - 4$

$-6x = -48$

$\dfrac{-6x}{-6} = \dfrac{-48}{-6}$

$x = 8$

Check:
$4 - 6(8) = -44$
$4 - 48 = -44$

---

18) $-12 - 4n = 36$

$-12 + 12 - 4n = 36 + 12$

$-4n = 48$

$\dfrac{-4n}{-4} = \dfrac{48}{-4}$

$n = -12$

Check:
$-12 - 4(-12) = 36$
$-12 + 48 = 36$
$36 = 36$

2.3 Solving Equations and Word Problems

Now that we have begun to learn the basic **skill** of solving equations, we can use that skill to solve word problems. The key to solving word problems is to **translate** words into mathematical symbols. In Section 1.4 we were introduced to some of the words used in the translations and this section will build on that skill.

Method for Solving Word Problems

The word problems in this section will be solved by using the following four step procedure:

1. Choose a **variable** to represent the unknown number.
2. Use that variable to **translate** the English sentence into a mathematical equation.
3. **Solve** the resulting equation.
4. Write a **sentence** that answers the question.

◆**Example 1** Five more than twice a number is 23. Find the number.

Solution: Let x = the number   Choose a **variable**

$5 + 2x = 23$ **Translation**

$5 + 2x - 5 = 23 - 5$ **Solve:** Subtract **5** from each side

$2x = 18$ Produce **0** to isolate the x

$\dfrac{2x}{2} = \dfrac{18}{2}$ Divide each side by **2**

$\boxed{x = 9}$ Produce **1**x

$\boxed{\text{The number is 9.}}$ Write a **sentence**

◆**Example 2** Six times a number decreased by seven will be 41. Find the number.

Solution: Let a = the number   Choose a **variable**

$6a - 7 = 41$ **Translation**

$6a - 7 + 7 = 41 + 7$ **Solve:** Add **7** to each side

$6a = 48$ Produce **0**, isolate the a

$\dfrac{6a}{6} = \dfrac{48}{6}$ Divide each side by **6**

$\boxed{a = 8}$ Produce **1**a

$\boxed{\text{The number is 8.}}$ Write a **sentence**

◆**Example 3** .Three times a number increased by nine was 45. Find the number

Solution: Let n = the number   Choose a **variable**

$$3n + 9 = 45$$ **Translation**

$$3n + 9 - 9 = 45 - 9$$ **Solve:** Subtract 9 from each side

$$3n = 36$$ Produce 0, isolate the n

$$\frac{3n}{3} = \frac{36}{3}$$ Divide each side by 3

$$\boxed{n = 12}$$ Produce 1 n

$$\boxed{\text{The number is 12}}$$ Write **sentence**

◆**Example 4** Translate the following problem. Do <u>not</u> solve.

When the quotient of five times a number and four is decreased by eight the result is two. What is the number?

Let w = the number, $\frac{5w}{4} - 8 = 2$

◆**Example 5** Translate the following problem. Do <u>not</u> solve.

Ten more than five times a number will be four increased by three times the number.

Let x = the number, $10 + 5x = 4 + 3x$.

Name:_____   2.3 Problem Set

Solve and check each of the following:

1) $3a + 2 = 14$

$3a + 2 - 2 = 14 - 2$
$3a = 12$
$\frac{3}{3} \quad \frac{3}{3}$
$a = 4$

Check:
$3(4) + 2 = 14$
$12 + 2 = 14$

2) $2v - 5 = 21$

$2v - 5 + 5 = 21 + 5$
$2v = 26$
$\frac{2v}{2} = \frac{26}{2}$
$v = 13$

Check:
$2(13) - 5 = 21$
$26 - 5 = 21$

3) $\frac{a}{5} - 2 = 4$

$\frac{a}{5} - 2 + 2 = 4 + 2$
$5 \cdot \frac{a}{5} = 6 \cdot 5$
$a = 30$

Check:
$\frac{30}{5} - 2 = 4$
$6 - 2 = 4$

4) $3 + \frac{n}{2} = 9$

$3 - 3 + \frac{n}{2} = 9 - 3$
$2 \cdot \frac{n}{2} = 6 \cdot 2$
$x = 12$

Check:
$3 + \frac{12}{2} = 9$
$3 + 6 = 9$

5) $26 = 3a + 5$

$26 - 5 = 3a + 5 - 5$
$21 = 3a$
$\frac{21}{3} = \frac{3a}{3}$
$7 = a$

Check:
$26 = 3(7) + 5$
$26 = 21 + 5$
$26 = 26$

6) $\frac{x}{3} + 7 = 13$

$\frac{x}{3} + 7 - 7 = 13 - 7$
$3 \cdot \frac{x}{3} = 6 \cdot 3$
$x = 18$

Check:
$\frac{18}{3} + 7 = 13$
$6 + 7 = 13$

7) $25 = -3 + 4w$

$25 + 3 = -3 + 3 + 4w$
$28 = 4w$
$\frac{28}{4} = \frac{4w}{4}$
$7 = w$

Check:
$25 = -3 + 4(7)$
$25 = -3 + 28$
$25 = 25$

8) $5 + 6x = 17$

$\quad -5 \quad -5$
$6x = 12$
$\frac{6}{6} \quad \frac{6}{6}$
$x = 2$

Check:
$5 + 6(2) = 17$
$5 + 12 = 17$

9) $2 - \frac{n}{2} = -4$

$2 - \frac{n}{2} = -4 - 2$
$-\frac{n}{2} = -6 \cdot 2$
$-n = -12$
$n = 12$

Check:
$2 - \frac{12}{2} = -4$
$2 - 6 = -4$
$-4 = -4$

Solve the following word problems:

10) Five more than twice a number is 37. Find the number.

$5 + 2x = 37$
$5 - 5 + 2x = 37 - 5$
$2x = 32$
$\frac{2x}{2} = \frac{32}{2}$
$x = 16$

11) Six times a number diminished by seven is 41. What is the number?

$6x - 7 = 41$
$6x + 7 = +7$
$6x = 48$
$\frac{6x}{6} = \frac{48}{6}$
$x = 8$

12) Eight times a certain number increased by 30 will be 6. What is the number?

$8x + 30 = 6$
$8x + 30 - 30 = 6 - 30$
$8x = -24$
$\frac{8x}{8} = \frac{-24}{8}$
$x = -3$

13) Five less than nine times a number is 31. Find the number.

$9x - 5 = 31$
$ +5 +5$
$9x = 36$
$\frac{9x}{9} = \frac{36}{9}$
$x = 4$

14) When a certain number divided by six was increased by 3 the result was ten. What was the number?

$\frac{x}{6} + 3 = 10$
$\frac{x}{6} + 3 - 3 = 10 - 3$
$\frac{x}{6} = 7$
$x = 42$

15) A certain number is divided by four and the result decreased by two forming a new number whose value is five. Find the number.

$\frac{x}{4} - 2 = 5$
$\phantom{\frac{x}{4}} +2 +2$
$4 \cdot \frac{x}{4} = 7 \cdot 4$
$x = 28$

2.4 Combining Like Terms in an Equation and Solving Equations with Variables on Both Sides of the Equal Sign

We will continue to build the skill of solving equations. In this section you will have to combine like terms in order to solve the equations. Also, you will learn to solve equations with variables located on both sides of the equal sign.

Combining Like Terms in an Equation

When like terms appear on the same side of an equation, it is very important to use the **skill** of **combining like terms** to simplify before attempting to undo the equation. You have already practiced combining like terms in Section 2.1. You will now use that skill in solving more complicated equations.

♦**Example 1** Solve $3x + 2 + 4x = 30$

$7x + 2 = 30$	Combine like terms $3x$ and $4x$
$7x + 2 - 2 = 30 - 2$	**Subtract 2** from each side
$7x = 28$	Produce **0**
$\dfrac{7x}{7} = \dfrac{28}{7}$	**Divide** each side by **7**
$\boxed{x = 4}$	Produce $1x$

Undoing Equations with Variables on Both Sides by Producing Zero and One

When variables appear on both sides of an equation, a mathematician will use the **property** of inverse operations to isolate the variable(s) on one side of the equation. As always, there are **decisions** to be made. The decisions will involve how to best proceed to **undo** the equation in a step-by-step manner until the variable stands alone.

The steps to undoing an equation with variables on both sides are:

1. **Eliminate the variable term on one side** of the equation by adding the opposite of the variable term to produce 0. In the following examples, we will eliminate the variables on the right side, but remember you can eliminate the variables from either side. The important thing to remember is get **all of your variables on one side.**

2. **Combine like terms whenever possible.** Examples of "like terms" are terms with constants only and terms that include the same variable(s) raised to the same degree.

3. Use an **inverse operation** to produce zero on the side where you have isolated the variables. Then use an inverse operation to produce **one** as the **coefficient of the variable.**

◆**Example 2** $5x + 6 - 2x = 17 + x + 1$ Note the variable "x" appears on both sides of the equation

$3x + 6 = x + 18$ **Combine like terms** $5x$ and $-2x$ (on left) and the like terms **17** and **1** (on right)

$3x + 6 - x = x + 18 - x$ **Eliminate variable** on right side

$2x + 6 = 18$ **Combine like terms**

$2x + 6 - 6 = 18 - 6$ **Undo** the **6 on left** by inverse operation. **Subtract 6** from each side

$2x = 12$ **Combine like terms**

$\dfrac{2x}{2} = \dfrac{12}{2}$ **Divide** each side by **2** to produce **1x**

$\boxed{x = 6}$

◆**Example 3** $5w + 8 = 2w - 7$ Note the variable "w" appears on both sides of the equation

$5w + 8 - 2w = 2w - 7 - 2w$ To eliminate the variable term from the right side **subtract 2w** from each side

$3w + 8 = 0 - 7$ Combine **5w** with **−2w** to produce **3w** on the left side

$3w + 8 - 8 = -7 - 8$ **Undo** the **8 on the left** by inverse operation. **Subtract 8** from each side

$3w + 0 = -15$ **Combine like terms**

$\dfrac{3w}{3} = \dfrac{-15}{3}$ **Divide** each side by **3** to produce **1w**.

$\boxed{w = -5}$

Helpful Hint 7: Don't bunch up your work! *Try to leave room on* **each** *side of the equation to add or subtract terms to both sides. Plan ahead. Start your work in the top left corner of the space given. Do not write so large that you cannot include all the necessary steps. If necessary, do the entire problem on another sheet, but be sure to label your work so that the teacher knows what problem you are working on the extra sheet.*

◆ **Example 4** Five more than six times a number is the same as twice the number decreased by three. Find the number.

    Solution: Let n = the number   **Choose a variable** to represent the number

$$5 + 6n = 2n - 3$$ Use variable to **translate**

$$5 + 6n - 2n = 2n - 3 - 2n$$ **Solve:** Subtract 2n from each side to **isolate** the variable on the left side

$$5 + 4n = -3$$ Combine **like terms** 6n and − 2n

$$5 + 4n - 5 = -3 - 5$$ Subtract **5** from each side

$$4n = -8$$ Combine **like terms** − 3 and − 5

$$\frac{4n}{4} = \frac{-8}{4}$$ Divide each side by **4**

$$\boxed{n = -2}$$ Produce 1n

$$\boxed{\text{The number is } -2.}$$ Write a **sentence**

Name:_____   2.4 Problem Set

Solve and check each of the following equations:

1) $5x - 2 + 2x = 40$

$7x - 2 = 40$
$7x - 2 + 2 = 40 + 2$
$7x = 42$
$\frac{7x}{7} = \frac{42}{7}$
$x = 6$

Check:
$5(6) - 2 + 2(6) = 40$
$30 - 2 + 12 = 40$
$30 + 10 = 40$
$40 = 40$

2) $7a + 8 - 3a = 56$

$4a + 8 = 56$
$4a + 8 - 8 = 56 - 8$
$4a = 48$
$\frac{4a}{4} = \frac{48}{4}$
$a = 12$

Check:
$7(12) + 8 - 3(12) = 56$
$84 + 8 - 36 = 56$
$84 - 28 = 56$
$56 = 56$

3) $9y - 3 = 4y + 32$

$-4y \quad -4y$
$5y - 3 = 32$
$+3 \quad +3$
$5y = 35$
$\frac{5y}{5} = \frac{35}{5}$
$y = 7$

Check:
$9(7) - 3 = 4(7) + 32$
$63 - 3 = 28 + 32$
$60 = 60$

4) $6r - 2 = 2r - 26$

$-2r \quad -2r$
$4r - 2 = -26$
$+2 \quad +2$
$4r = -24$
$\frac{4r}{4} = \frac{-24}{4}$
$r = -6$

Check:
$6(-6) - 2 = 2(-6) - 26$
$-36 - 2 = -12 - 26$
$-38 = -38$

5) $8n + 7 = 2n + 49$

$-2 \quad -2$
$6n + 7 = 49$
$-7 \quad -7$
$\frac{6n}{6} = \frac{42}{6}$
$n = 7$

Check:
$8(7) + 7 = 2(7) + 49$
$56 + 7 = 14 + 49$
$63 = 63$

6) $4x - 2 + 3x = 12 + 5x$

$7x - 2 = 12 + 5x$
$-5 \quad -5$
$2x - 2 = 12$
$+2 \quad +2$
$\frac{8x}{8} = \frac{14}{2}$
$x = 7$

Check:
$4(7) - 2 + 3(7) = 12 + 5(7)$
$28 - 2 + 21 = 12 + 35$
$47 = 47$

7) $9 - 3w = 5w + 57$

$+3 \quad +3$
$9 = 8w + 57$
$-57 \quad -57$
$-48 = 8w$
$\frac{-48}{8} = \frac{8w}{8}$
$-6 = w$

Check:
$9 - 3(-6) = 5(-6) + 57$
$9 + 18 = -30 + 57$
$27 = 27$

8) $3 - 2n = 38 + 3n$

$+2 \quad +2$
$3 = 38 + 5n$
$-38 \quad -38$
$\frac{-35}{5} = \frac{5n}{5}$
$-7 = n$

Check:
$3 - 2(-7) = 38 + 3(-7)$
$3 + 14 = 38 - 21$
$17 = 17$

9) $6x - 7 + x - 35 = 0$

$7x - 42 = 0$
$7x - 42 + 42 = 0 + 42$
$\frac{7x}{7} = \frac{42}{7}$
$x = 6$

Check:
$6(6) - 7 + 6 - 35 = 0$
$36 - 1 - 35 = 0$
$35 - 35 = 0$
$0 = 0$

2.4 Problem Set   Name:_____

10) $14 = 2x - 21 + 5x$

$-5 \quad -5$

$14 = -3x - 21$
$+21 +21$

$35 = -3x$
$-3 \quad -3$

$11\tfrac{2}{3} = x$

Check:

11) $18 = 7x + 30 - 4x$

Check:

12) $8a - 3 + 5a = 10a - 27$

Check:

Solve the following word problems:

13) Five times a certain number decreased by three is the same as twice the number increased by 18. Find the number.

$5c - 3 = 2x + 18$

14) When seven is added to nine times a certain number the result is the same as twice the number diminished by 35. What is the number?

$7 + 9c = 2c + 35$

15) Four more than six times a number is twice the number added to 32. What is the number?

$4 + 6x = 2x + 32$

16) Ten less than seven times a number will be four more than nine times the number. What is the number?

$10 - 7c = 4 + 9c$

2.5 Solving Equations and Word Problems Using the Distributive Property

We will continue building on the skill of solving equations. In this section we will use the distributive property learned in section 2.1 to solve equations and word problems.

REVIEW OF THE DISTRIBUTIVE PROPERTY

Rule: When a number is multiplied by a sum or difference of two numbers, the number outside the parenthesis is multiplied by each number inside the parenthesis. This property is called the distributive property.

Symbolically: $a(b+c) = ab + ac$ and $a(b-c) = ab - ac$.

◆**Example 1** $5(a + b) = 5a + 5b$ 5 is multiplied by each of a and b or one can say 5 is distributed

◆**Example 2** $8(2s - 5) = 8(2s) - 8(5)$ 8 is distributed
$= 16s - 40$

◆**Example 3** $-4(3v - 2a) = -4(3v) - (-4)(2a)$ –4 is distributed
$= -12v + 8a$

◆**Example 4** The following equation involves use of the distributive property.

Solve: $8(x + 3) = 6(x - 2)$

$8(x) + 8(3) = 6(x) - 6(2)$ Distributive property

$8x + 24 - 6x = 6x - 12 - 6x$ Subtract **6x** from each side

$2x + 24 = -12$ Combine terms, produce **0**

$2x + 24 - 24 = -12 - 24$ Subtract **24** from each side

$2x = -36$ Produce **0**, combine terms, work down

$\dfrac{2x}{2} = \dfrac{-36}{2}$ Divide each side by **2**

$\boxed{x = -18}$ Produce **1x**

Check: $8[(-18) + 3]\ ?\ 6[(-18) - 2]$? used to indicate that we are not sure yet if sides are equal

$8[-15]\ ?\ 6[-20]$

$\boxed{-120 = -120}$ True

Further work on Translating Words into Mathematical Symbols and Use of Grouping Symbols such as Parenthesis and Brackets.

Rule: When a constant or variable is multiplied by a sum or difference of two numbers, the sum or difference should be placed in parenthesis when translating the words into mathematical symbols

Example: The expression in words: Five times the sum of a number and six

Translation into symbols: Let x = the number,

5(x + 6)

Example: The expression in words: Four times the difference of twice the number and nine

Translation into symbols: Let x = the number,

4 (2x − 9)

Example: The expression in words: Three more than six times the sum of four times a number and five

Translation into symbols: Let n = the number,

3 + 6(4n + 5)

Example: The expression in words: Six more than four times the difference of five times a number and nine is ten.

Translation: Let y = the number,

6 + 4(5y − 9) = 10

◆ **Example 5** Find the number such that seven times the difference of the number and three is six less than four times the number

Solution: Let x = the number	Choose a **variable** for the unknown number
7 (x − 3) = 4x − 6	**Translation**
7x − 21 = 4x − 6	**Solve:** Distribute the **7**
7x − 21 − 4x = 4x − 6 − 4x	Subtract **4x** from each side to eliminate the variable from the right side
3x − 21 = −6	Combine **7x** with **−4x** (like terms)
3x − 21 + 21 = −6 + 21	Add **21** to each side to eliminate the constant term from the left side
3x = 15	Combine like terms on the right side
$\frac{3x}{3} = \frac{15}{3}$	Divide each side by **3** to produce **1x**
$\boxed{x = 5}$	
$\boxed{\text{The number is 5.}}$	Write a **sentence**.

◆Example 6

Six times the sum of twice a number and four is eight more than ten times the number. What is the number?

Solution:	
Let n = the number	Choose a **variable** for the unknown number
6(2n + 4) = 8 + 10n	**Translation**
12n + 24 = 8 + 10n	**Solve:** Distribute the **6**
12n + 24 − **10n** = 8 + 10n − **10 n**	Subtract **10n** from each side to eliminate the variable from the right side
2n + 24 = 8	Combine **12n** and **−10n** (like terms)
2n + 24 − **24** = 8 − **24**	Subtract **24** from each side to eliminate the constant on the left side
2n = −16	Combine the constants (like terms) on the right side
$\dfrac{2n}{2} = \dfrac{-16}{2}$	Divide each side by **2** to produce **1n**
$\boxed{n = -8}$	
$\boxed{\text{The number is } -8.}$	Write a **sentence**.

Name:_____   2.5 Problem Set

Solve each of the following problems. Show the check on the problems indicated.

1) $3(x+3) = 8(x-2)$	2) $-2(2x+1) = 7(x-5)$	3) $-32 = -8(2w+4)$
Check:	Check:	Check:
4) $16 - 3(2b+5) = 3b - 17$	5) $11(2x-3) - 17x = 3x - 13$	6) $0 = 4(6-3c)$
7) $7y + 3(5y-2) = 38$	8) $10 - 2(6-3x) = -8$	9) $-27 = -9 + 3(x+6)$

Translate each of the following using x to represent the number. **DO NOT SOLVE.**

10) Eight times the sum of a number and six is nine more than five times the number.

11) Three more than four times the difference of a number and ten is twice the number increased by 58.

12) Seven more than three times the difference of a number and six will be six times the number.

13) Five less than seven times the difference of a number and eight was 86.

14) Three times the sum of twice a number and seven is six more than five times the number.

15) Four more than five times the sum of six times a number and seven will be ten times the number.

16) One less than four times the difference of three times a number and eight is six times the number.

17) Eight less than six times the sum of nine times a number and two was nine more than twice the number.

18) Twice a number deceased by four times the sum of nine times a number and seven is five greater than eight times the difference of three times a number and two.

19) Six more than five times a number increased by seven times the difference of four times the number and three will be ten more than three times the sum of the number and twelve.

Solve the following word problems:

20) Eight more than five times the sum of a number and six is 58 increased by three times the number. Find the number.

21) Seven more than four times the difference of a number and five will be nine more than twice the number. What is the number?

22) What number increased by four times the difference of twice the number and seven is the same as two added to five times the sum of the number and two?

23) When six times the difference of five and four times a certain number is increased by twice the sum of eight times the number and nine the result is zero. Find the number.

2.6 Chapter Review

In this chapter, you learned to:

1.. Solve Equations.
2. Solve number type word problems.

Properties of Real Numbers   [2.1]

The following statements are true for each real number a, b and c.

Commutative	$a + b = b + a$ and $a \cdot b = b \cdot a$
Associative	$(a + b) + c = a + (b + c)$ and $(a \cdot b) \cdot c = a \cdot (b \cdot c)$
Distributive	$a(b + c) = ab + ac$
Identity elements	$a + 0 = a$ and $a \cdot 1 = a$
Inverse operations	$a + (-a) = 0$ and $a \cdot \frac{1}{a} = 1$, $a \neq 0$

Properties of Equality   [2.2]

The following statements are true for each real number a, b and c.

Addition	If $a = b$ then $a + c = b + c$
Subtraction	If $a = b$ then $a - c = b - c$
Multiplication	If $a = b$ then $a \cdot c = b \cdot c$
Division	If $a = b$ then $\frac{a}{c} = \frac{b}{c}$, $c \neq 0$

Steps Used to Solve an Equation [2.2, 2.4]

1. **Combine like terms** whenever possible.
2. **Isolate the variable term** on one side of the equation (**produce zero**)
3. Produce **variable** term with **coefficient of one**.

Solving Number Type Word Problems   [2.3]

Use the 4 step method to solve number type word problems.

1. Assign a **variable** to represent the unknown number
2. Write an **equation** translating the words into mathematical symbols
3. **Solve** the equation
4. Write a **sentence** that answers the question of the problem

Chapter 2: Equations and Word Problems

◆**Example 1** Eight more than five times the difference of three times a number and four is twice the sum of five times a number and nine. What is the number?

Let **n** = the number	Assign a **variable** to represent the unknown number
$8 + 5(3n - 4) = 2(5n + 9)$	**Translation**
$8 + \mathbf{15n - 20} = 10n + 18$	**Solve: Distribute the 5**
$\mathbf{-12} + 15n = 10n + 18$	Combine **like terms** 8 and -20
$-12 + 15n \mathbf{-10n} = 10n + 18 \mathbf{-10n}$	**Subtract 10n** from each side
$5n - 12 = 18$	Combine **like terms**
$5n - 12 + 12 = 18 + 12$	**Add 12** to each side
$5n = 30$	Combine **like terms**
$\dfrac{5n}{5} = \dfrac{30}{5}$	Divide each side **by 5**
$\boxed{n = 6}$	Produce **1n**
$\boxed{\text{The number is 6}}$	Write a **sentence**

◆**Example 2** When four times the sum of six times a number and five is diminished by seven the result is the same as five times the difference of four times the number and three. Find the number?

Let x = the number	Assign a **variable**
$4(6x + 5) - 7 = 5(4x - 3)$	**Translation**
$24x + 20 - 7 = 20x - 15$	**Solve:** Distribute
$24x + 13 = 20x - 15$	Combine like terms
$24x + 13 - 20x = 20x - 15 - 20x$	Subtract 20x from each side
$4x + 13 = -15$	Combine like terms
$4x + 13 - 13 = -15 - 13$	Subtract 13 from each side
$4x = -28$	Combine like terms
$\dfrac{4x}{4} = \dfrac{-28}{4}$	Divide each side by 4
$\boxed{x = -7}$	Produce 1x
$\boxed{\text{The number is } -7.}$	Write a **sentence.**

Name:_____ 2.6 Chapter Review Test

Solve the following equations in the usual manner:

1) $5 + 2a = 17$	2) $3 + 5(x - 4) = 2x + 1$
3) $4 - 6(2x + 3) = 34$	4) $-6 + 5(2x - 4) = 4 + 3(2x - 6)$
5) $3m - 4(2m - 7) = m - 44$	6) $5y + 9(y - 4) - 6(y + 8) = 4$

Evaluate each of the following if $a = -2$ $b = 3$ $x = -5$ $w = 6$

7) $a + b[w - x]$	8) $bx - aw$	9) $[w + a]^2 + [bw - a)$

Solve each of the following word problems:

10) Six more than twice a number is five times the sum of the number and three. Find the number.

11) When eight is added to three times the sum of a number and four the result is five times the number decreased by six. What is the number?

12) If seven times a certain number is diminished by twice the difference of the number and six the result is twice the number decreased by three. Find the number.

13) Five times the difference of twice a number and eight is the same as 31 more than seven times the sum of the number and four. What is the number?

14) Three less than six times the sum of a number and five will be nine. Find the number.

15) Seven times the sum of twice a number and three is nine more than three times the sum of four times the number and six. Find the number.

16) Five times the difference of three times a number and seven is one more than six times the sum of twice a number and five. What is the number?

17) Eight is the same as two increased by six times the difference of a number and seven. Find the number.

CHAPTER 3
Applications of Equations

In this chapter, you will

1. *Review solving of linear equations.*
2. *Use the 5 step process to solve word problems.*

3.1 Solving Word Problems

In Section 2.3 we learned a 4 step process for solving number type word problems. Now, we will add one more step in the process to solve word problems dealing with a broader spectrum of word problems. In Chapter Two, we were asked only to find one unknown number. In this Chapter, we will be asked to solve word problems that ask for more than one unknown number. For example, we may be asked to determine the ages of two different people or the cost of two different items. In general we will expand our ability to solve problems that are more complex.

In order to conquer these more difficult word problems we must **think like a mathematician**. This means we must **organize information** in a structured manner in order to **see patterns, make decisions** and use **skills** to solve the problem.

Follow these 5 steps:

1. Set up **categories** of information given.
2. Assign **numbers** to the categories.
3. Write an **equation**.
4. **Solve** the equation.
5. Write the answer in the form of a **sentence**.

Step 1: Set Up Categories of Information Given

By reading the problem, determine what type of numbers are being discussed. We will discover that number words like **cost of, length of, height of, age of** are being used. These are what we call **categories** of information. In order to properly **organize the information** of the problem, we must clearly establish in our minds the nature of these number-type words. Step #1 is to write the categories down on paper.

Step 2: Assign Numbers to the Categories

Once we have listed the categories in an orderly manner, we will reread the problem and translate the words into mathematical symbols and numbers and assign these symbols and numbers to the proper categories. This is a difficult step, because at this point we will have to admit that we don't know something. We will have to use a variable, like x or n, to represent what we don't know. Here we will notice a **pattern** among the numbers.

Step 3: Write an Equation

Another rereading of the problem will reveal how the numbers relate to each other. We must make a **decision** as to how the facts of the problem relate to each other. The translation of the relationship of these facts using mathematical symbols is the equation.

Step 4: Solve the Equation

Now we have the opportunity to employ the **skill** of solving an equation that we learned in Chapter 2. Be sure to check the solution by substituting the value found for the variable in the equation. We must also ask ourselves if the answer found is a reasonable one for the facts of the problem. Use common sense!

Step 5: Write a Sentence

Now that we have solved the equation, we must write a sentence that answers the question in the word problem. It is a good idea to **underline the question or the demand** to assure that the sentence is the response called for in the problem. In other words, we should ask ourselves if we made the right **decision** relative to the problem posed.

◆**Example 1** Tarzan is six years older than twice the age of Boy. When three times Tarzan's age is added to five times Boy's age the result is 205 years. Find the sum of their ages.

1) The **categories** here deal with **age** type numbers.

   = **age** of Boy

   = **age** of Tarzan

2) Assign **numbers** to the above two categories

   let x = age of Boy. Start with what you don't know.

   $2x + 6$ = age of Tarzan. Relate age of Tarzan to age of Boy. [Remember that the first sentence above says that "Tarzan is **six more than twice** the age of Boy."]

3) Form an **equation** that relates the information given.

   $3(2x + 6) + 5x = 205$. Make a **decision**. [Remember that the second sentence above says that "when **three times Tarzan's age is added to five times Boy's age the result is 205**."]

4) **Solve** the equation. [You already know how to do this!]

$3(2x + 6) + 5x = 205$	
$6x + 18 + 5x = 205$	Distribute the **3**
$11x + 18 = 205$	Combine **6x** and **5x**
$11x + 18 - 18 = 205 - 18$	Subtract **18** from each side
$11x = 187$	Produce **0**, work down
$\dfrac{11x}{11} = \dfrac{187}{11}$	Divide each side by **11**
$\boxed{x = 17}$	Produce **1** as the coefficient of x, work down

5) Write a **sentence** that answers the question or the demand. In this problem there is no question but rather a demand to find the sum of Tarzan's and Boy's ages. Remember to underline the question or demand. When we solved for x, we found the age of Boy to be 17. We know that Tarzan's age = $2x + 6$, so

$2x + 6 = 2(17) + 6$

$= 34 + 6$

$2x + 6 = 40$

So now that we know that Boy is 17 and Tarzan is 40, we can respond to the demand.

The sum of their ages is 57.

Helpful Hint 8: *In Example 1 above, the categories appear as step #1. When you actually work a word problem, **the categories will not appear as a separate step.** Rather, you will write the categories on the right with an equal sign in front of them and then assign the numbers or a variable to the left of the = sign.*

◆**Example 2**   Wally bought a desk and a lamp to improve his home study environment. The desk cost $15 more than four times the cost of the lamp. The total purchase price of these two items was $175. If he had bought a computer it would cost six times as much as the desk. What would be the cost of the computer?

Solution:

= **cost** of lamp	Establish **categories** with number type words
= **cost** of desk	
Let c = cost of lamp	Choose a variable (**number**) to represent the **cost** you do not know.
$4c + 15$ = cost of desk	Relate cost of desk to cost of lamp. "**The desk cost $15 more than 4 times the cost of lamp.**"
$c + 4c + 15 = 175$	Form an **equation** that relates the numbers. "**The total purchase price of these two items was $175.**"
$5c + 15 = 175$	**Solve** the equation
$5c + 15 - 15 = 175 - 15$	
$5c = 160$	
$\dfrac{5c}{5} = \dfrac{160}{5}$	
$\boxed{c = 32}$	Remember that you let c = cost of lamp

We have found the cost of the lamp to be $32. The desk is represented by $4c + 15$ so by substituting $c = 32$ into the expression, $4c + 15$ we have

$4c + 15 = 4(32) + 15$

$\qquad\quad\; = 128 + 15$

$\qquad\quad\; = 143$

So the cost of the desk is $143. However the **question** asks for the cost of a computer which costs **six times** as much as the desk. Now we must multiply the cost of the desk ($143) by 6, and we have the cost of the computer: $6(\$143) = \$858$.

The computer would cost $858.   Write a **sentence**

Chapter 3: Applications of Equations

♦ **Example 3** The larger of two numbers is three more than five times the smaller. If their sum is 45, <u>what is the larger number</u>?

Solution:

= **smaller** number	Establish **categories** using number type words
= **larger** number	
Let **n** = smaller number	Choose a variable to represent the **number** that you do not know, in this case the smaller
5n + 3 = larger number	Relate larger number to smaller. **"The larger number is 3 more than 5 times the smaller"**
n + 5n + 3 = 45	Form an **equation** that relates the numbers. Recall that **"their sum is 45"**
6n + 3 = 45	**Solve** the equation
6n + 3 − 3 = 45 − 3	
6n = 42	
$\dfrac{6n}{6} = \dfrac{42}{6}$	
$\boxed{n = 7}$	

Remember we let n = the **smaller** number. The <u>question</u> asks for the **larger number**. We let 5n + 3 represent the larger number, so we have:

$$5n + 3 = 5(7) + 3$$
$$= 35 + 3$$
$$5n + 3 = 38$$

$\boxed{\text{The larger number is 38.}}$	Write a **sentence**

Use the 5 step method to solve the following problems:

1. The larger of two numbers is five more than three times the smaller number. If their sum is 29, find the numbers.

2. The age of Boris is seven years more than twice the age of Ivan. If the sum of their ages is 52, find the age of Boris.

3. Marta recently paid $290 for a bicycle and a helmet. She paid two dollars more than five times as much for the bike as the helmet. What was the price of the bike?

4. Abdul bought two books and three computer disks. Each book cost twice as much as each disk (including tax). If his total bill was exactly $21, how much did he pay for each book?

5. In a certain class, there are eight more than twice as many women as men. If the class has 47 students, how many women are in the class?

6. Find two numbers such that the larger is five less than three times the smaller. When the larger is multiplied by four and then decreased by three times the smaller the result is 25.

7. The weight of Tarzan is 30 pounds less than twice the weight of Madonna. Four times Madonna's weight added to three times Tarzan's weight is 960 pounds. Find the weight of Ferd Burfil who weighs five pounds less than Tarzan.

8. Gilda is seven years less than four times the age of Ivan. When Ivan's age is multiplied by five and increased by twice the age of Gilda, the result is 64. How old is Gilda?

9. The larger of two numbers is five less than three times the smaller. Seven times the larger diminished by twice the smaller is 41. Find the larger number.

10. For her exercise workout, Olga rode her bike a distance of three miles more than twice the distance that she jogged. If the total distance she traveled was 21 miles, how far did she ride her bike?

11. Mortimer is three years older than twice the age of Alfonso. If the sum of their ages is 51, find the age of Mortimer.

12. The cost of the speakers for Corine's new stereo system was $45 more than three times the cost of the receiver. What did her speakers cost if the total system cost $929?

3.2 More Word Problems

In Section 3.1, we used the 5 step process to solve word problems involving ages of one or more persons, costs of different items, distance traveled, and so forth. In this section, we will practice the 5 step process again. We will also work new types of word problems.

Remember that whenever we solve a word problem, we must think like a mathematician. Categories and numbers help **organize information.** To assign numbers to the categories we must recognize the **patterns** given in the problem. The equation formation involves **making decisions.** Solving the equation uses several **skills** and the writing of a sentence involves **decision making** again. The problems get harder, but the strategy remains the same.

---

*Helpful Hint 9: Draw a picture! On certain types of problems, it is a good idea to draw a picture to help **organize information** and to **see patterns**. The saying "a picture is worth a thousand words" reminds us that a sketch helps us visualize the information given in a problem. Depending on the problem, the picture might be a line segment, rectangle, triangle, circle or a combination of these basic shapes. The idea is to use as many of your senses as possible to help organize your thoughts and to understand the problem.*

◆**Example 1** A rectangular garden is located against a back fence with the longer side against the fence. Its length is six feet less than twice its width and its perimeter is 60 feet. <u>Find the cost of fencing the three remaining sides</u> if fencing costs $4 per linear foot.

Solution: By drawing the figure, we can better visualize the information given and the question. In solving this problem, we recall the perimeter formula for a rectangle:
$$P = 2L + 2W.$$

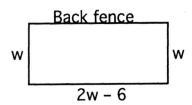

w = the **width** of the rectangle	**Categories** and **numbers**
2w − 6 = the **length** of the rectangle	The length is 6 feet less than twice its width.
2(2w − 6) + 2w = 60	**Form equation** using perimeter formula for rectangle, 2L + 2W = P
4w − 12 + 2w = 60	**Solve** the equation
6w − 12 = 60	
6w − 12 + 12 = 60 + 12	
6w = 72	
$\frac{6w}{6} = \frac{72}{6}$	
$\boxed{w = 12}$ and $\boxed{2w − 6 = 18}$	
width length	

We have solved for w, so we can say that the width is 12 feet and the length is 18 feet. However the problem asks for the <u>cost of fencing the three remaining sides</u>. By looking at the picture, we visualize its solution. Since the back fence is already in place, we only need to buy fence for the remaining **three** sides. By adding up the three sides we get 42 feet, and since each foot costs $4, we multiply $4(42) and get a total cost of $168.

| The cost of the three sides of the fence is $168. | Write a **sentence** |

◆**Example 2** One side of a triangle is twice another side. The third side is eight feet less than the sum of the first two sides. If the perimeter is 70 feet, <u>how long is the longest side</u>?

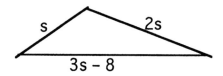

Solution:

Let s = **length** of the shortest side	**Categories** and **numbers**
2s = **length** of second side	"One side is twice another side."
3s – 8 = **length** of the third side	"Third side is 8 feet less than the sum of the first two sides." (s + 2s = 3s)
s + 2s + 3s – 8 = 70	Form an **equation:** sum of 3 sides (perimeter) is 70
6s – 8 = 70	**Solve** the equation
6s – 8 + 8 = 70 + 8	
6s = **78**	
$\frac{6s}{6} = \frac{78}{6}$	

$\boxed{s = 13}$ and $\boxed{3(13) - 8 = 31}$

shortest side   longest side

| The longest side is 31 feet. | Write a **sentence** |

Name:_____   3.2 Problem Set   93

Use the 5 step method to solve the following word problems:

1. A triangle has a perimeter of 78 meters. Two sides have the same length and the third side is six meters less than the sum of the other two sides. Find the length of the largest side.

2. The longest side of a triangle is 4 cm more than three times the shortest side. The third side is twice the shortest side. If the perimeter is 46 cm, find the length of the largest side.

3. Carol recently bought a book, a computer disk, and a pogo stick, paying a total of $126. She paid two dollars more than three times as much for the book as the disk, while the pogo stick cost twice as much as the book. Find the cost of the pogo stick.

4. Connie used her new $450 bike to travel the six miles to the math party. During the second hour of the party she worked eight more than twice as many problems as Kim, but their combined total of 47 problems was the highest pair at the party. If the party ended at 9 PM sharp, how many problems did Connie work?

5. The floor of Batman's costume storage closet is nine feet more than three times its width. If the perimeter is 138 feet, what is its length?

6. Fannie Farkel selected a rectangular frame to place the picture of Ferd Burfil. The length of the frame was 8 cm less than twice its width. Find the dimensions of the frame if its perimeter was 68 cm.

7. Ivan slept two hours less than Carlos, while Boris slept as much as Ivan and Carlos combined. If these guys slept a total of 32 hours, how many hours did Boris sleep?

8. The larger of two numbers is six less than twice the smaller. When five times the smaller is increased by three times the larger, the result is 114. Find the sum of these two numbers.

3.3 Word Problems Involving Sum of Two Numbers

There are many situations involving <u>two</u> unknown quantities in which their sum is known. We may buy a hamburger and a drink for a total of $ 4. We may take a 500 mile trip, part by car, part by bicycle. We may know that the sum of the ages of a mother and daughter total 50 years. In each of these examples, a sum of <u>two</u> quantities is known but the first and second quantities are unknown. We will now look at the **pattern** involved and use this pattern to solve problems.

When the sum of two numbers is known as well as the value of one of the two numbers, the other number can be found

We know from learning number facts, that if the sum of two numbers is 10 and one of the numbers is 8, the other number is 2. This may seem like a trivial question, but what is the pattern that is used to get the answer? The **pattern** is that we subtracted the known quantity "8" from the sum "10": $10 - 8 = 2$

Example: The sum of two numbers is 30 and one number is 24. What is the other?

$30 - 24 = 6$. The other number is 6.

Example: The sum of two numbers is 75 and one number is 25. What is the other?

$75 - 25 = 50$. The other number is 50.

What is the pattern? $\boxed{\text{Subtract the known quantity from the sum}}$

Use the above pattern to set up categories for word problems, in which you are given the sum of two numbers, but neither of the numbers is known

This section includes word problems where the sum of two numbers is given, but we are not told the value of <u>either</u> of the numbers. The first step in solving a word problem is to set up the **categories**. Where the categories are two numbers whose sum is given, use the pattern set forth above.

Pattern: When you know the sum of two numbers,

let x = one number

the sum $- x$ = the other number.

Example: The sum of two numbers equals 15. What are the two numbers?

Let x = one number.

$15 - x$ = the other number.

Example: The sum of two numbers is 100. What are the two numbers?

Let x = one number

$100 - x$ = the other number.

Example: In a carton of a dozen eggs, some are broken; some are not. How many of each are there?

Let y = number of broken eggs

12 − y = number of unbroken eggs

◆**Example 1** Find two numbers whose sum is 35 such that six times the smaller diminished by twice the larger is two.

Solution: Let x = smaller number   **Categories** and **numbers**

35 − x = larger number

6x − 2(35 − x) = 2   Form **equation.** Translate, "six times the smaller diminished by twice the larger is two".

6x − 70 + 2x = 2   **Solve** the equation

8x − 70 = 2

8x − 70 + 70 = 2 + 70

8x = 72

$$\frac{8x}{8} = \frac{72}{8}$$

x = 9   Smaller number   and   35 − 9 = 26   Larger number

The two numbers are 9 and 26.   Answer question with a **sentence**

◆**Example 2** Hector purchased a pencil and a notebook for a total cost of $5. When his math class saw Hector using these items to organize his math notes, many rushed to the bookstore to buy the same items for themselves spending a total of $105 on 23 pencils and 18 notebooks. Find the cost of each notebook?

Solution: Let x = **cost** of a pencil   **Categories** and **numbers**

5 − x = **cost** of a notebook

23x + 18(5 − x) = 105   Form **equation** using "spending a total of $105 on pencils and 18 notebooks"

23x + 90 − 18x = 105   **Solve** the equation

5x + 90 = 105

5x + 90 − 90 = 105 − 90

5x = 15

$$\frac{5x}{5} = \frac{15}{5}$$

x = 3   Cost of pencil   and   5 − 3 = 2   Cost of notebook

The notebooks cost $2 each.   Write a **sentence**.

Use the 5 step process to solve these problems:

1) A computer and printer can be purchased in the bookstore for $1025. Fannie Farkel decided to buy five printers and two computers paying a total of $3175. Find the cost of each.

2) The sum of two numbers is 20. Five times the larger increased by twice the smaller will be 79. Find the smaller number.

3) The age of Pat is three times the age of Ben, while Etar is five years younger than Pat. If the sum of their ages is 79, find the age of Etar.

4) Juanita spent $20 for a movie and dinner. Six times the cost of the movie decreased by three times the dinner cost would be $3. Find the cost of the movie.

5) Ferd Burfil has a rectangular back yard. He would like to build a fence on the two sides and the back. The back is 12 feet less than twice the length of each side. The total length of the fence is 156 feet. How long is the back fence?

6) A class has 36 students. Four times the number of girls added to five times the number of boys is exactly 159. How many boys are in the class?

7) A sack has 100 marbles, some are red, some are blue. Four times the number of red marbles diminished by twice the number of blue marbles is 286. How many of each color marble are in the sack?

8) On a recent shopping spree, Lisa spent three times as much for her rocket powered roller blade skates as she did for her gold plated yo yo. The laser guided pogo stick cost six dollars less than the yo yo. Her total expenditure was $839. Find the cost of the roller blade skates.

Solve each of the following equations:

9) $6 + 2(3x - 5) = 10x - 8$	10) $5 - 3(4x - 7) = 6 - 7x$	11) $8 + 5(3x - 8) = 7x$
Check:	Check:	Check:
12) $-4 - 4(2x + 5) + 5x = x$	13) $a - 2 = -9 - 2(3a + 7)$	14) $-n + 3(2n + 5) = 0$
Check:	Check:	Check:

Evaluate each of the following if $a = -3$ $n = 2$ $x = -5$ $v = 8$

15) $a^2n - x^2v$	16) $\dfrac{nxv}{x - a}$	17) $ax^2 - n[v + x]$

Name: _____

Solve the following problems using the 5 step method. (For help refer to previous lessons)

1. Kareem bought a Mickey Mouse pencil and a blue cap with a propeller on top, paying a total of $19. The cap cost four dollars more than twice the cost of the pencil.. Find the cost of the cap.

2. The length of a rectangle is six feet more than three times its width. If the perimeter is 68 feet, find its length.

3. The lowest grade on a recent math exam was a grade of C. (Students learned to think like a mathematician.) There were twice as many A grades as B grades and the number of C grades was nine less than the number of A grades. How many grades were A if the class had 36 students?

4. Lucia bought 3 CDs and 5 blank video tapes for a total of $59. Each CD cost one dollar more than three times the cost of each video tape. Find the cost of each CD.

5. The combined age of Amanda and her Mom is 71 years. Her Mom is seven years more than three times the age of Amanda. How old is the Mother?

6. Two sides of a triangle have the same length, while the longest side is seven meters less than four times the length each of the equal sides. If the total perimeter of the triangle is 77 meters, find the length of the longest side.

7. The sum of two numbers is 63. When twice the smaller is added to five times the larger the result is 276. Find the two numbers.

8. The combined weight of Laurel, Hardy and Ferd Burfil is 531 pounds. Hardy weighs 23 pounds less than twice the weight of Laurel, while Ferd weighs 14 pounds more than Laurel. Find the weight of Ferd Burfil.

3.5 Chapter Review

In this chapter, you learned to:

1. Solve Equations.
2. Use the 5 step process to solve word problems.

The overall goal of this course is to think like a mathematician. This means we need to:

1. **Organize** information
2. Look for and use **patterns**
3. Make **decisions**
4. Develop **skills**

Solving Word Problems   [3.1]

In this chapter we have been thinking like mathematicians by working word problems using a 5 step process which is summarized below:

1. Set up **categories** of information.
2. Assign **numbers** to the categories.
3. Write an **equation**.
4. **Solve** the equation.
5. Write a **sentence**.

◆**Example 1**   The combined weight of Laurel and Hardy is 349 pounds. If five times Laurel's weight decreased by twice Hardy's weight would be 303 pounds, what is the weight of each?

Solution: Let x = weight of Laurel	**Categories** and **numbers**
$349 - x$ = weight of Hardy	Sum minus the other number
$5x - 2(349 - x) = 303$	Form **equation** by translation of "five times Laurel's weight decreased by twice Hardy's weight would be 303 pounds."
$5x - 698 + 2x = 303$	**Solve** equation. Distributive property
$7x - 698 = 303$	Combine like terms
$7x - 698 + 698 = 303 + 698$	Add 698 to each side to produce 0
$7x = 1001$	Combine like terms
$\dfrac{7x}{7} = \dfrac{1001}{7}$	Divide each side by 7 to produce 1 as coefficient of x

$\boxed{x = 143 \quad \text{weight of Laurel}}$ and $\boxed{349 - 143 = 206 \quad \text{weight of Hardy}}$

$\boxed{\text{Laurel weighs 143 pounds and Hardy weighs 206 pounds.}}$ Write a **sentence**.

Solve the following equations in the usual manner.

1) $5 + 2a = 47$	2) $4 + 5(x - 4) = 2x + 35$	3) $7 - 4(3w + 9) = 31$
4) $9 + 5(2x - 4) = 4 + 3(2x - 9)$	5) $8m - 4(6m - 5) = m - 31$	6) $8y + 4(y - 7) - 6(y + 8) = 34$

Evaluate each of the following if $a = -2$ $b = 3$ $x = -5$ $w = 4$

7) $a + b[x - w]$	8) $bx - aw$	9) $[w - a]^2 + x^2 a^3$

Solve the following word problems using the 5-step method.

10. The combined weight of Boris and Ivan is 360 pounds. If Boris weighs six pounds more than twice the weight of Ivan, how much does Boris weigh?

11. The larger of two numbers is 8 less than three times the smaller number. Twice the larger number diminished by four times the smaller is 28. Find the sum of the two numbers.

12. Tarzan is eight years more than three times the age of Ferd Burfill. If the sum of their ages is 84, find the age of Batman who is five years younger than Tarzan.

13. Find two numbers whose sum is 73 and when twice the smaller is added to three times the larger the result is 193.

14. When Steve Martin was 38 years old he bought a rectangular lot at a total cost of $ 240,000. The lot was four feet less than three times its width. If the perimeter was 432 feet, what was the length.

15. Fannie Farkel drove 42 miles to a sale in her 1994 Honda. Driving with care she averaged 36 MPG. At the sale she purchased a hat, pants and gloves to accent her Fall wardrobe. She paid $18 more for the pants than the gloves and twice as much for the hat as the pants. How much did she pay for the hat if the total purchase price was $162.

16. Wanda is three years less than five times as old as Gilda. If twice the age of Wanda is decreased by four times the age of Gilda, the result would be exactly 78 years. How old will Wanda be in two short years ?

17. Boris ate five more tacos than Igor, while Big Al ate three times as many as Boris. How many tacos did Boris eat if together they consumed a total of 45 tacos ?

CHAPTER 4
Fractions

In this chapter, you will learn to:
1. *Perform the operations of addition, subtraction, multiplication and division on fractions.*
2. *Solve word problems involving fractions.*

4.1 Introduction to Fractions

Before proceeding to the operations on fractions, we need to review the types of fractions, factoring, prime numbers, and raising and reducing fractions.

What are common fractions?

A common fraction is a number expressed as the quotient of two integers. The word quotient means division, so we will use a division bar to express the quotient. Examples of common fractions are

$$\frac{2}{3}, \quad \frac{5}{6}, \quad \frac{9}{4} \quad \text{and} \quad \frac{7}{3}$$

It will be easier to talk about fractions if we know the names for their parts. In the fraction,

$\frac{a}{b}$ — the top number is called the **numerator**
— the division or fraction bar
— the bottom number is called the **denominator** (b≠0)

There are two types of common fractions: proper and improper.

What is a proper fraction?

A fraction is **proper** if the numerator is smaller than the denominator. Examples of proper fractions are

$$\frac{2}{3}, \quad \frac{7}{8} \quad \text{and} \quad \frac{5}{6}$$

The fraction $\frac{2}{3}$ is read as 2 divided by 3 or two-thirds.

The fraction $\frac{7}{8}$ is read as 7 divided by 8 or seven-eighths.

The fraction $\frac{5}{6}$ is read as 5 divided by six or five-sixths.

We can visualize a fraction as a part of a whole. Consider the circle below as a whole.

We can visualize $\frac{3}{4}$ by dividing the whole into 4 equal parts and shading in 3 parts.

We can visualize $\frac{2}{3}$ as

What is an Improper Fraction?

A fraction is **improper** if the numerator is larger than the denominator. Examples of improper fractions are

$$\frac{5}{4}, \quad \frac{7}{2} \quad \text{and} \quad \frac{8}{3}$$

Again, we can visualize the fraction by using a geometric shape.

To visualize $\frac{5}{4}$ we can divide two rectangles into fourths (4 equal parts) because we see that $\frac{4}{4}$ equals one whole rectangle and that we have an additional one-fourth. Since the picture represents $\frac{4}{4}$ and $\frac{1}{4}$ more we say that

$$\frac{4}{4} + \frac{1}{4} = \frac{5}{4}$$

| $\frac{1}{4}$ | $\frac{1}{4}$ | $\frac{1}{4}$ | $\frac{1}{4}$ | + | $\frac{1}{4}$ | | |

The improper fraction $\frac{5}{3}$ can be visualized by using 2 circles, each divided into thirds (3 equal parts). As you can see from these examples, the denominator is the number of equal parts into which the figure is divided.

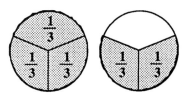

What is a Mixed Number?

Besides the common fractions, there is another class of fractions called mixed numbers. A **mixed number** is the sum of a whole number and a proper fraction. Examples of mixed numbers are

$$2\frac{3}{4}, \quad 5\frac{7}{9} \quad \text{and} \quad 8\frac{2}{5}$$

You know from your pie eating days that one and a half apple pies look like this:

1 pie + $\frac{1}{2}$ pie = $1\frac{1}{2}$ pie

$1\frac{1}{4}$ pies looks like

1 pie + $\frac{1}{4}$ pie = $1\frac{1}{4}$ pie

Looking back to improper fractions and looking at the above picture, we can see that $1\frac{1}{4}$ can also be written as $\frac{5}{4}$ because $\frac{4}{4} = 1$ so $\frac{4}{4} + \frac{1}{4} = \frac{5}{4} = 1\frac{1}{4}$ or

$\frac{5}{4}$ pies is the same as $1\frac{1}{4}$ pies. (Aunt Sally went on and cut the whole pie into fourths in case we want more!)

How to Use the Number One in Fractions.

Whenever the numerator is the same as the denominator, the value of the fraction is one. Symbolically we write, $\frac{a}{a} = 1$, if $a \neq 0$. Examples of fractions that equal one are $\quad \frac{3}{3} = 1 \qquad \frac{6}{6} = 1 \qquad$ and $\qquad \frac{-5}{-5} = 1$

Preparation for Reducing Fractions to Lowest Terms.

Before we learn the process of reducing fractions, we need to know how to find the greatest common factor of a set of numbers and we need to become acquainted with a set of numbers called the prime numbers. A mathematician tries to do things in the simplest and most organized manner. Since lower terms (smaller integers) are usually easier to work with, we will want to be skilled in reducing fractions to their lowest terms. For example:

$\frac{6}{8}\quad$ can be reduced to $\quad\frac{3}{4}\quad$ so we say that $\quad\frac{6}{8} = \frac{3}{4}$

$\frac{9}{15}\quad$ can be reduced to $\quad\frac{3}{5}\quad$ so we say that $\quad\frac{9}{15} = \frac{3}{5}$

A fraction may be reduced if its numerator and denominator have a common factor (other than 1). The factors of a given number are numbers that divide the given number evenly (no remainder).

For example, the factors of 6 are 1, 2, 3 and 6 because each of these integers divide evenly into 6. The factors of 8 are 1, 2, 4, and 8 because each of these integers divide evenly into 8. The only factors of 7 are 7 and 1. When the only factors of an integer are the integer itself and one, we call the integer **prime.**

To reduce a fraction to its lowest terms, we should first find the greatest common factor of the numerator and denominator and use the concept of 1 to cancel. This process will be explained after we have discussed how to find the greatest common factor.

The Greatest Common Factor (GCF)

To find the greatest common factor (GCF) of two or more integers, first find the factors of each and then determine which common factor is the greatest.

◆**Example 1** Find the greatest common factor (GCF) of 12 and 18

The factors of 12 are 1, 2, 3, 4, **6** and 12

The factors of 18 are 1, 2, 3, **6**, 9 and 18

The common factors are 2, 3 and 6 but the greatest common factor is 6.

◆**Example 2** Find the GCF of 24 and 36

The factors of 24 are 1, 2, 3, 4, 6, 8, **12** and 24.

The factors of 36 are 1, 2, 3, 4, 6, 9, **12**, 18 and 36.

Again, there are many common factors but the GCF is 12.

Using the GCF will lead us to our first method of reducing fractions.

Method 1: Reducing Fractions Using the GCF

1. Find the **GCF** of the numerator and the denominator.
2. **Factor** each of the numerator and denominator using the GCF as one of the factors.
3. **Cancel** the GCF using the identity property of the number 1.
4. Write the product of the remaining factors of the numerator as the numerator of reduced fraction.
5. Write the product of the remaining factors of the denominator as the denominator of the reduced fraction.

◆**Example 3** Reduce $\frac{15}{20}$ to lowest terms.

$$\frac{15}{20} = \frac{5 \cdot 3}{5 \cdot 4} \quad \text{GCF is 5}$$

$$= \frac{3}{4} \quad \text{Cancel, } \frac{5}{5} = 1$$

◆**Example 4** Reduce $\frac{36}{60}$ to lowest terms.

$$\frac{36}{60} = \frac{12 \cdot 3}{12 \cdot 5} \quad \text{GCF is 12}$$

$$= \frac{3}{5} \quad \text{Cancel, } \frac{12}{12} = 1$$

Prime Factors

Another method of reducing fractions is to find all the prime factors of the numerator and denominator and then to cancel common factors. Prime factors must always be prime integers. A **prime integer** is any integer greater than 1 whose only factors are the number itself and 1. We could also say that a prime integer is any integer greater than 1 that has exactly two divisors, itself and 1. A prime factor is a **divisor** of another integer if it divides the other integer evenly (without a remainder). An integer greater than 1 that is not prime is a **composite** integer.

The integer 2 is prime, because its only factors are 2 and 1.

The integer 3 is prime, because its only factors are 3 and 1.

The integer 4 is composite, because 2 is a factor as well as 4 and 1.

The integer 5 is prime because its only factors are 5 and 1.

The integer 6 is composite because 2 and 3 are factors as well as 6 and 1.

A partial listing of the prime integers could be $\{2, 3, 5, 7, 11, 13, 17, 19, 23, 29, \ldots\}$

One method of finding the prime factors of an integer is to use a **prime tree**. First divide the given integer by the first prime 2. If 2 divides the given integer evenly (without a remainder), then 2 is a prime factor of the given integer. If two divides into the given number evenly, then divide the remainder by 2 again and continue so long as the remainder is even. If 2 does not divide into the given number or the remainder evenly, then divide by the next prime 3. If 3 divides the given integer or the remainder evenly, then 3 is a prime factor. Continue in same manner using the remaining primes as the divisor. The prime factors then will be the complete set of integers that divide the given integer. The following examples illustrate the process. The prime factorization is written as the product of all of the prime factors used to make up the product. Notice that some of the factors appear more than once.

◆**Example 5**   Find the prime factors of 24 using a prime tree.

$24 = 2 \cdot 12$ 2 is a prime factor. The remainder is 12.

$24 = 2 \cdot 2 \cdot 6$ 2 is a prime factor. The remainder is 6.

$24 = 2 \cdot 2 \cdot 2 \cdot 3$ 2 is a prime factor. 3 is the remainder and is a prime so you know that 3 is the last prime factor.

> The prime factors of 24 are 2, 2, 2 and 3.

◆**Example 6**   Find the prime factors of 90 using a prime tree.

$90 = 2 \cdot 45$

$90 = 2 \cdot 3 \cdot 15$

$90 = 2 \cdot 3 \cdot 3 \cdot 5$

> The prime factors of 90 are 2, 3, 3, 5.

Method 2: Reducing Fractions Using Prime Factors

We are now ready for another method of reducing fractions.

1. **Factor** the numerator and denominator into primes.
2. **Cancel** common factors using the identity property of one.
3. **Multiply** remaining factors in the numerator and write as numerator of reduced fraction.
4. **Multiply** remaining factors in the denominator and write as denominator of reduced fraction.

◆**Example 7**   Reduce $\dfrac{60}{126}$ to lowest terms.

$$\frac{60}{126} = \frac{2 \cdot 2 \cdot 3 \cdot 5}{2 \cdot 3 \cdot 3 \cdot 7}$$

$$\boxed{= \frac{10}{21}}$$

◆**Example 8**   Reduce $\dfrac{45}{54}$ to lowest terms.

$$\dfrac{45}{54} = \dfrac{3 \cdot 3 \cdot 5}{3 \cdot 3 \cdot 3 \cdot 2}$$

$$\boxed{= \dfrac{5}{6}}$$

Building to a Higher Terms

There are times when we want to write a fraction as an equivalent fraction having a higher denominator. This is particularly useful when we add or subtract fractions. We will again use the identity property of 1, by multiplying the numerator and denominator of the fraction by the same number. This concept is sometimes referred to as the fundamental theorem of fractions.

> Fundamental Theorem of Fractions: If $\dfrac{a}{b}$ is a fraction and if $b \neq 0$ and $c \neq 0$, then
> $$\dfrac{a}{b} = \dfrac{a \cdot c}{b \cdot c}$$

◆**Example 9**   Write the fraction $\dfrac{2}{3}$ as an equivalent fraction having a denominator of 12.

We first ask "What can we multiply the denominator 3 by to obtain the desired denominator of 12?" The answer is $3 \cdot 4 = 12$. So we multiply the fraction by 1 in the form of $\dfrac{4}{4}$.

$$\dfrac{2}{3} = \dfrac{2 \cdot 4}{3 \cdot 4}$$

$$= \dfrac{8}{12}$$

So we say that $\dfrac{2}{3} = \dfrac{8}{12}$

◆**Example 10**   Write the fraction $\dfrac{3}{7}$ as an equivalent fraction having a denominator of 35.

$$\dfrac{3}{7} = \dfrac{3 \cdot 5}{7 \cdot 5}$$

$$= \dfrac{15}{35}$$

So we say $\dfrac{3}{7} = \dfrac{15}{35}$

Name: _____   4.1 Problem Set   117

Represent each of the following fractions visually by first dividing the figure into the correct number of equal parts and then shading the parts necessary to describe the given fraction.

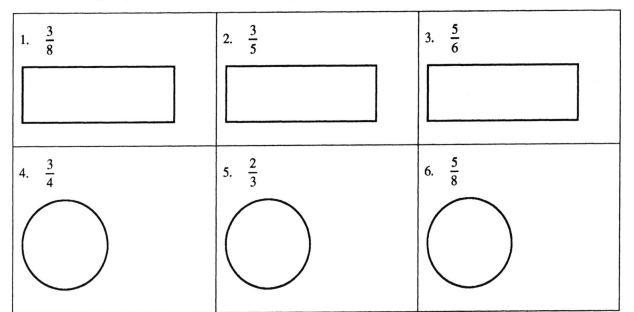

1. $\dfrac{3}{8}$

2. $\dfrac{3}{5}$

3. $\dfrac{5}{6}$

4. $\dfrac{3}{4}$

5. $\dfrac{2}{3}$

6. $\dfrac{5}{8}$

Find the GCF (Greatest Common Factor) of each of the following.

7. 6 and 8	8. 12 and 20	9. 15 and 35

Reduce each of the following using GCF method

10. $\dfrac{20}{24}$	11. $\dfrac{15}{18}$	12. $\dfrac{15}{35}$

13. $\dfrac{16}{40}$	14. $\dfrac{30}{12}$	15. $\dfrac{40}{32}$

4.1 Problem Set   Name:_____

Use a prime tree to find the prime factors of each of the following.

16. 63	17. 180	18. 120

Reduce each of the following using prime factors

19. $\frac{12}{15}$	20. $\frac{18}{24}$	21. $\frac{15}{18}$
22. $\frac{42}{35}$	23. $\frac{44}{28}$	24. $\frac{14}{70}$

Build each of the following fractions to an equivalent fraction with indicated higher term.

25. $\frac{3}{4} = \frac{}{24}$	26. $\frac{5}{8} = \frac{}{72}$	27. $\frac{3}{5} = \frac{}{40}$
28. $\frac{7}{3} = \frac{}{9}$	29. $\frac{8}{5} = \frac{}{40}$	30. $\frac{3}{10} = \frac{}{60}$

4.2 Operations on Common Fractions I

Section 4.1 introduced the basic concepts of common fractions with skill practice in reducing fractions to lowest terms and raising fractions to higher terms. Now you will use some of these skills as you multiply and divide common fractions. Work with addition and subtraction of common fractions will be limited to fractions with common denominators. As always, organization of information is very important, so be neat, work down and enjoy.

The Operation of Multiplication on Common Fractions

To multiply common fractions:

1. Multiply the **numerators** and write the product as the numerator of the answer.

2. Multiply the **denominators** and write the product as the denominator of the answer.

Symbolically: $\dfrac{a}{b} \cdot \dfrac{c}{d} = \dfrac{a \cdot c}{b \cdot d}$ if $b \neq 0$ and $d \neq 0$

◆**Example 1**   Find the product. $\dfrac{5}{8} \cdot \dfrac{3}{4}$

$$\dfrac{5}{8} \cdot \dfrac{3}{4} = \dfrac{5 \cdot 3}{8 \cdot 4}$$

$$\boxed{= \dfrac{15}{32}}$$

◆**Example 2**   Find the product. $\dfrac{3}{5} \cdot \dfrac{6}{7}$

$$\dfrac{3}{5} \cdot \dfrac{6}{7} = \dfrac{3 \cdot 6}{5 \cdot 7}$$

$$\boxed{= \dfrac{18}{35}}$$

Multiplication of Common Fractions with Cancellation

Some multiplication problems have fractions that can be canceled because at least one pair of numerators and denominators have a common factor. When a common factor is involved:

1. **Factor** the numerator and denominator of each fraction into primes.
2. **Cancel** (produce the identity 1) factors common to the numerators and denominators.
3. **Multiply** the remaining **numerators** and write the product as the numerator of the answer.
4. **Multiply** the remaining **denominators** and write the product as the denominator of the answer.

◆**Example 3** Find the product. $\dfrac{5}{6} \cdot \dfrac{18}{35}$

$$\dfrac{5}{6} \cdot \dfrac{18}{35} = \dfrac{5}{2 \cdot 3} \cdot \dfrac{2 \cdot 3 \cdot 3}{5 \cdot 7}$$ Factor, cancel

$$= \dfrac{3}{7}$$

◆**Example 4** Find the product. $\dfrac{3}{8} \cdot \dfrac{12}{5}$

$$\dfrac{3}{8} \cdot \dfrac{12}{5} = \dfrac{3}{2 \cdot 2 \cdot 2} \cdot \dfrac{2 \cdot 2 \cdot 3}{5}$$ Factor, cancel

$$= \dfrac{9}{10}$$

◆**Example 5** Find the product. $\dfrac{49}{15} \cdot \dfrac{9}{14}$

$$\dfrac{49}{15} \cdot \dfrac{9}{14} = \dfrac{7 \cdot 7}{3 \cdot 5} \cdot \dfrac{3 \cdot 3}{7 \cdot 2}$$ Factor, cancel

$$= \dfrac{21}{10}$$

Division with Common Fractions

As discussed in Section 2.1, division is the inverse of multiplication. We will use this inverse concept to establish the pattern for division. For a common fraction, we obtain the inverse, also called the reciprocal, by interchanging the numerator and the denominator.

Symbolically, the reciprocal of the fraction, $\frac{a}{b}$ is the fraction $\frac{b}{a}$ (a ≠ 0 and b ≠ 0).

For example,

the reciprocal of $\frac{3}{2}$ is $\frac{2}{3}$

the reciprocal of $\frac{3}{4}$ is $\frac{4}{3}$

the reciprocal of 6 is $\frac{1}{6}$

To divide two common fractions, we do the following:

1. Multiply the first fraction by the reciprocal of the second fraction.
2. Use the procedures outlined in the multiplication of fractions process.

◆**Example 6** Find the quotient. $\frac{2}{3} \div \frac{5}{6}$

$$\frac{2}{3} \div \frac{5}{6} = \frac{2}{3} \cdot \frac{6}{5} \quad \text{Multiply first fraction by reciprocal of second}$$

$$= \frac{2}{3} \cdot \frac{2 \cdot 3}{5}$$

$$\boxed{= \frac{4}{5}}$$

◆**Example 7** Find the quotient. $\frac{5}{8} \div \frac{3}{16}$

$$\frac{5}{8} \div \frac{3}{16} = \frac{5}{8} \cdot \frac{16}{3} \quad \text{Multiply first fraction by reciprocal of second}$$

$$= \frac{5}{2 \cdot 2 \cdot 2} \cdot \frac{2 \cdot 2 \cdot 2 \cdot 2}{3}$$

$$\boxed{= \frac{10}{3}}$$

Addition or Subtraction of Common Fractions with a Common Denominator

When two or more fractions have the same denominator, we say they have a common denominator.

The fractions $\frac{2}{5}$ and $\frac{1}{5}$ both have a denominator of 5 so 5 is the common denominator of the two fractions.

Fractions that have common denominators are considered like terms and therefore we can use the same **pattern** developed in Chapter 2, namely combining like terms. For example,

$$\frac{2}{5} + \frac{1}{5} = \frac{3}{5}$$ means 2 fifths added to 1 fifth is 3 fifths.

Use the same pattern of combining like terms for subtraction of fractions with a common denominator.

$$\frac{6}{7} - \frac{2}{7} = \frac{4}{7}$$ means subtracting 2 sevenths from 6 sevenths is 4 sevenths

The rules for adding or subtracting fractions with a common denominator are:

1. To **add** two fractions with a common denominator, **add the numerators** and place the sum over the common denominator. Reduce answer if possible.

2. To **subtract** two fractions with the same denominator, **subtract the numerators** and place the difference over the common denominator. Reduce answer if possible.

◆**Example 8** Find the sum. $\frac{2}{9} + \frac{4}{9}$

$$\frac{2}{9} + \frac{4}{9} = \frac{6}{9}$$

$$\boxed{= \frac{2}{3}}$$

◆**Example 9** Find the difference. $\frac{11}{12} - \frac{5}{12}$

$$\frac{11}{12} - \frac{5}{12} = \frac{6}{12}$$

$$\boxed{= \frac{1}{2}}$$

Mixing Operations

When more than one operation appears in an expression, remember the order of operations rule (Please Elect My Dear Aunt Sally) and perform the operations in the following order:

1. **Parenthesis** or other grouping
2. **Exponents**
3. **Multiplication** or **Division** (whichever comes first working left to right)
4. **Addition** or **Subtraction** (whichever comes first working left to right)

◆**Example 10** Perform the indicated operation. $\dfrac{4}{15} + \dfrac{2}{3} \cdot \dfrac{4}{5}$

$$\dfrac{4}{15} + \dfrac{2}{3} \cdot \dfrac{4}{5} = \dfrac{4}{15} + \dfrac{8}{15}$$

$$= \dfrac{12}{15}$$

$$= \dfrac{2 \cdot 2 \cdot 3}{3 \cdot 5}$$

$$\boxed{= \dfrac{4}{5}}$$

◆**Example 11** Perform the indicated operation. $\dfrac{7}{8} \cdot \dfrac{2}{5} \div \dfrac{14}{15}$

$$\dfrac{7}{8} \cdot \dfrac{2}{5} \div \dfrac{14}{15} = \dfrac{7}{2 \cdot 2 \cdot 2} \cdot \dfrac{2}{5} \div \dfrac{14}{15}$$

$$= \dfrac{7}{20} \cdot \dfrac{15}{14}$$

$$= \dfrac{7}{2 \cdot 2 \cdot 5} \cdot \dfrac{3 \cdot 5}{2 \cdot 7}$$

$$\boxed{= \dfrac{3}{8}}$$

◆ **Example 12** Perform the indicated operation. $\dfrac{5}{18} + \left(\dfrac{2}{3}\right)^2 \cdot \dfrac{7}{8}$

$$\dfrac{5}{18} + \left(\dfrac{2}{3}\right)^2 \cdot \dfrac{7}{8} = \dfrac{5}{18} + \dfrac{4}{9} \cdot \dfrac{7}{8}$$

$$= \dfrac{5}{18} + \dfrac{2 \cdot 2}{3 \cdot 3} \cdot \dfrac{7}{2 \cdot 2 \cdot 2}$$

$$= \dfrac{5}{18} + \dfrac{7}{18}$$

$$= \dfrac{12}{18}$$

$$\boxed{= \dfrac{2}{3}}$$

Name:_____

4.2 Problem Set

Find the product.

1) $\dfrac{5}{7} \cdot \dfrac{2}{3}$	2) $\dfrac{7}{8} \cdot \dfrac{2}{5}$	3) $\dfrac{5}{6} \cdot \dfrac{3}{5}$
4) $\dfrac{7}{15} \cdot \dfrac{5}{7}$	5) $\dfrac{5}{3} \cdot \dfrac{9}{10}$	6) $\dfrac{4}{3} \cdot \dfrac{15}{8}$
7) $\dfrac{5}{8} \cdot \dfrac{12}{7}$	8) $\dfrac{16}{3} \cdot \dfrac{5}{8}$	9) $\dfrac{4}{9} \cdot \dfrac{27}{8}$

Find the quotient.

10) $\dfrac{2}{3} \div \dfrac{5}{7}$	11) $\dfrac{5}{6} \div \dfrac{10}{3}$	12) $\dfrac{7}{8} \div \dfrac{5}{12}$
13) $\dfrac{4}{5} \div \dfrac{16}{3}$	14) $\dfrac{5}{8} \div \dfrac{10}{7}$	15) $\dfrac{9}{16} \div \dfrac{15}{4}$

4.2 Problem Set Name:_____

Find the sum.

16) $\dfrac{2}{3} + \dfrac{5}{3}$

17) $\dfrac{5}{8} + \dfrac{3}{8}$

18) $\dfrac{7}{8} + \dfrac{3}{8}$

19) $\dfrac{5}{6} + \dfrac{11}{6}$

20) $\dfrac{2}{15} + \dfrac{7}{15}$

21) $\dfrac{5}{8} + \dfrac{7}{8}$

Find the difference.

22) $\dfrac{8}{9} - \dfrac{2}{9}$

23) $\dfrac{7}{12} - \dfrac{5}{12}$

24) $\dfrac{8}{15} - \dfrac{2}{15}$

25) $\dfrac{11}{6} - \dfrac{7}{6}$

26) $\dfrac{17}{12} - \dfrac{5}{12}$

27) $\dfrac{19}{14} - \dfrac{3}{14}$

Name:_____

Perform the indicated operation(s).

28) $\dfrac{5}{6} \cdot \dfrac{2}{3}$

29) $\dfrac{8}{21} + \dfrac{2}{3} \cdot \dfrac{5}{7}$

30) $\dfrac{7}{8} \div \dfrac{5}{4}$

31) $\dfrac{47}{48} - \dfrac{5}{6} \cdot \dfrac{7}{8}$

32) $\dfrac{9}{5} \div \dfrac{7}{10} \cdot \dfrac{25}{9}$

33) $\dfrac{4}{9} \cdot \dfrac{7}{8} \div \dfrac{7}{3}$

Use a factor tree to factor each of the following into primes.

34) 150

35) 240

36) 200

Find the GCF (greatest common factor) of each of the following:

37) 30 and 45

38) 12 and 42

39) 16 and 20

4.3 Operations on Common Fractions II

In this section you will learn to add and subtract common fractions that <u>do not</u> have a common denominator. Since you must have a common denominator to add or subtract fractions, our attention will be directed to developing the skill of finding the least common denominator, and then changing the given fractions to equivalent fractions having the least common denominator. To find the least common denominator we must develop a strategy. This means that we will **organize** information, look for **patterns** and make **decisions.**

Least Common Denominator (LCD)

The least common denominator (LCD) of a group of fractions is the same as the least common multiple (LCM) of the set of integers making up the denominators. We will begin by learning how to find the LCM.

The **multiples** of an integer are the integers obtained by multiplying the given integers by each integers from the set of whole numbers 1, 2, 3, 4, 5, 6, ... For example,

The multiples of 3 are 3•1, 3•2, 3•3, 3•4, 3•5,...= 3, 6, 9, 12, 15,....

The multiples of 5 are 5•1, 5•2, 5•3, 5•4, 5•5, ... = 5, 10, 15, 20, 25,....

The multiples of 10 are 10, 20, 30, 40, 50, 60,....

The **least common multiple** (LCM) of a set of integers is the smallest integers that is a multiple of each integer in the set.

◆**Example 1** Find the LCM of 12, 15 and 20.

The multiples of 12 are:	12, 24, 36, 48, **60,** 72, 84, 96, 108, **120,** ...
The multiples of 15 are:	15, 30, 45, **60,** 75, 90, 105, **120,** 135,...
The multiples of 20 are:	20, 40, **60,** 80, 100, **120,** 140, 160, ...

In the above example, both 60 and 120 are common multiples as well as many other common multiples not shown here, but the least common multiple is 60.

Let us now seek a more organized approach to finding an LCM by examining the prime factors of each of the above integers.

The prime factors of 12 are: 2•2•3

The prime factors of 15 are: 3•5

The prime factors of 20 are: 2•2•5

Since the given integers 12, 15 and 20 must be factors of the LCM, we know that the factors of the first integer 12, namely 2•2•3, are a part of the LCM. Next we look for other prime factors that must also be a part of the LCM. The second integer 15 must be a part, so 3 and 5 must be factors. But since we already have the 3 as a factor, all we need is the 5. Finally 20 must be a part, so 2•2•5 must be factors, but these already appear in the factors of the previous numbers. So using the prime factors, the LCM = 2•2•3•5 = **60.**

130 Chapter 4 : Fractions

In summary, to find the LCM of two or more integers:

1. Factor each integer into primes.
2. Use each factor the most it is used in any one integer to form the LCM.

◆**Example 2**   Find the LCM of 18, 20 and 54.

$$18 = 2 \cdot 3 \cdot 3$$ Factor each into primes

$$20 = 2 \cdot 2 \cdot 5$$ "

$$54 = 2 \cdot 3 \cdot 3 \cdot 3$$ "

$$\text{LCM} = 2 \cdot 2 \cdot 3 \cdot 3 \cdot 3$$ Use each factor the most it is used in any one number

The LCM is 108.

◆**Example 3**   Find the LCM of 14, 35 and 40.

$$14 = 2 \cdot 7$$ Factor into primes

$$35 = 5 \cdot 7$$

$$40 = 2 \cdot 2 \cdot 2 \cdot 5$$

$$\text{LCM} = 2 \cdot 7 \cdot 5 \cdot 2 \cdot 2 \text{ or } 2 \cdot 2 \cdot 2 \cdot 5 \cdot 7$$ Use each factor the most

The LCM is 280.

Addition and Subtraction with Different Denominators

As stated above, the LCM is also the LCD (least common denominator). We will now use our skill of finding the LCM to write equivalent fractions having a common denominator. We will also be using the fundamental theorem of fractions (Section 4.1), which states:

$$\frac{a}{b} = \frac{a}{b} \cdot \frac{c}{c}, \text{ if } b \neq 0 \text{ and } c \neq o$$

To add or subtract two or more fractions with different denominators:

1. Find LCD for the denominators using the LCM method stated above.
2. Change each fraction to an equivalent fraction having the least common denominator. (Multiply by 1 in the form that would obtain the common denominator.)
3. Use the rule for addition or subtraction of fractions with the same denominator.

◆**Example 4**   Find the sum: $\dfrac{5}{6} + \dfrac{3}{4}$

First find the LCD of 6 and 4 which is 12.

$$\frac{5}{6}+\frac{3}{4}=\frac{5}{6}\left(\frac{2}{2}\right)+\frac{3}{4}\left(\frac{3}{3}\right)$$

$$=\frac{10}{12}+\frac{9}{12}$$

$$\boxed{=\frac{19}{12}}$$

◆**Example 5** Find the difference: $\dfrac{7}{9}-\dfrac{1}{6}$

First find the LCD which is 18.

$$\frac{7}{9}-\frac{1}{6}=\frac{7}{9}\left(\frac{2}{2}\right)-\frac{1}{6}\left(\frac{3}{3}\right)$$

$$=\frac{14}{18}-\frac{3}{18}$$

$$\boxed{=\frac{11}{18}}$$

◆**Example 6** Perform the indicated operation: $\dfrac{5}{6}-\dfrac{1}{3}$

The LCD is 6

$$\frac{5}{6}-\frac{1}{3}=\frac{5}{6}-\frac{1}{3}\left(\frac{2}{2}\right)$$

$$=\frac{5}{6}-\frac{2}{6}$$

$$=\frac{3}{6}$$

$$\boxed{=\frac{1}{2}}$$

◆**Example 7** Perform the indicated operation: $\dfrac{5}{6}+\dfrac{1}{2}-\dfrac{1}{4}$

The LCD is 12

$$\frac{5}{6}+\frac{1}{2}-\frac{1}{4}=\frac{5}{6}\left(\frac{2}{2}\right)+\frac{1}{2}\left(\frac{6}{6}\right)-\frac{1}{4}\left(\frac{3}{3}\right)$$

$$=\frac{10}{12}+\frac{6}{12}-\frac{3}{12}$$

$$= \frac{16}{12} - \frac{3}{12}$$

$$\boxed{= \frac{13}{12}}$$

◆**Example 8** Perform the indicated operation: $\frac{5}{6} - \frac{3}{8} + \frac{1}{4}$

The LCD is 24

$$\frac{5}{6} - \frac{3}{8} + \frac{1}{4} = \frac{5}{6}\left(\frac{4}{4}\right) - \frac{3}{8}\left(\frac{3}{3}\right) + \frac{1}{4}\left(\frac{6}{6}\right)$$

$$= \frac{20}{24} - \frac{9}{24} + \frac{6}{24}$$

$$= \frac{11}{24} + \frac{6}{24}$$

$$\boxed{= \frac{17}{24}}$$

◆**Example 9** Perform the indicated operation: $\frac{12}{5} \div \frac{3}{4} - \frac{3}{10}$

More than one operation. Remember: Please Elect My Dear Aunt Sally.

$$\frac{12}{5} \div \frac{3}{4} - \frac{3}{10} = \frac{12}{5} \cdot \frac{4}{3} - \frac{7}{10}$$

$$= \frac{2 \cdot 2 \cdot 3}{5} \cdot \frac{2 \cdot 2}{3} - \frac{7}{10}$$

$$= \frac{16}{5} - \frac{7}{10}$$

$$= = \frac{16}{5}\left(\frac{2}{2}\right) - \frac{7}{10}$$

$$= \frac{32}{10} - \frac{7}{10}$$

$$= \frac{25}{10}$$

$$\boxed{= \frac{5}{2}}$$

Perform the indicated operations. Reduce answers to lowest terms.

1) $\dfrac{3}{16} + \dfrac{7}{16}$

2) $\dfrac{5}{8} + \dfrac{1}{8}$

3) $\dfrac{5}{6} - \dfrac{1}{6}$

4) $\dfrac{7}{9} - \dfrac{2}{9}$

5) $\dfrac{7}{8} + \dfrac{5}{8} - \dfrac{3}{8}$

6) $\dfrac{11}{12} - \dfrac{5}{12} + \dfrac{4}{12}$

7) $\dfrac{3}{8} + \dfrac{5}{6}$

8) $\dfrac{3}{4} - \dfrac{1}{3}$

9) $\dfrac{7}{9} - \dfrac{2}{3}$

10) $\dfrac{5}{6} + \dfrac{2}{3} - \dfrac{1}{2}$

11) $\dfrac{3}{4} - \dfrac{2}{3} + \dfrac{5}{6}$

12) $\dfrac{7}{8} + \dfrac{5}{6} - \dfrac{1}{4}$

Perform the indicated operations.

13) $\dfrac{5}{6} \cdot \dfrac{2}{3}$

14) $\dfrac{7}{8} \cdot \dfrac{4}{5}$

15) $\dfrac{2}{5} \cdot \dfrac{8}{9}$

16) $\dfrac{3}{4} + \dfrac{2}{3} \cdot \dfrac{3}{5}$

17) $\dfrac{3}{4} \cdot \dfrac{2}{5} + \dfrac{3}{2}$

18) $\dfrac{5}{4} + \dfrac{15}{7} \div \dfrac{10}{21}$

19) $\dfrac{9}{2} - \dfrac{2}{3} + \dfrac{3}{4}$

20) $\dfrac{2}{5} + \dfrac{3}{8} \cdot \dfrac{4}{9}$

21) $\dfrac{15}{4} - \dfrac{2}{3} \div \dfrac{4}{9}$

22) $\dfrac{25}{4} \div \dfrac{15}{8} + \dfrac{5}{6}$

23) $\dfrac{5}{9} \div \dfrac{5}{12} - \dfrac{1}{6}$

24) $\dfrac{15}{2} - \dfrac{4}{9} \div \dfrac{5}{18}$

4.4 Operations on Mixed Numbers

In these first few sections we have concentrated on common fractions. We will now consider another type of fraction called a mixed number. A **mixed number** is the sum of a whole number and a proper fraction. In this section we will learn to perform the four basic operations on mixed numbers. Remember to be neat, work down and have fun.

Convert a Mixed Number to an Improper Fraction

The mixed number $2\frac{1}{3}$ means that the whole number 2 is added to the proper fraction $\frac{1}{3}$. Since $2 = \frac{6}{3}$ we can write $2\frac{1}{3} = \frac{6}{3} + \frac{1}{3}$ or $\frac{7}{3}$

Another way to do the above problem would be to multiply the whole number (2) by the denominator (3) and add this product to the numerator (1). The result is placed over the denominator (3) of the fractional part. These steps are shown below.

$$2\frac{1}{3} = \frac{2 \cdot 3 + 1}{3} = \frac{7}{3}$$

◆**Example 1** Convert $3\frac{4}{5}$ to an improper fraction.

$$3\frac{4}{5} = \frac{3 \cdot 5 + 4}{5} \qquad \text{Denominator (5) times whole number (3) plus}$$
$$\text{numerator (4), all over the denominator (5)}$$

$$= \boxed{\frac{19}{4}}$$

◆**Example 2** Convert $6\frac{2}{7}$ to an improper fraction.

$$6\frac{2}{7} = \frac{7 \cdot 6 + 2}{7} \qquad \text{Denominator (7) times whole number (6) plus numerator (2)}$$
$$\text{all over the denominator (7)}$$

$$= \boxed{\frac{44}{7}}$$

Convert an Improper Fraction to a Mixed Number

To convert from an improper fraction into a mixed number, divide the numerator by the denominator. The quotient will be the whole number part. The fractional part is the remainder over the given denominator. For example,

the improper fraction $\frac{8}{5}$ is the same as $1\frac{3}{5}$ because $8 \div 5 = 1$ with a remainder of 3.

136 Chapter 4: Fractions

◆**Example 3** Convert $\frac{23}{6}$ to a mixed number.

$$\frac{23}{6} = 3\frac{5}{6} \qquad \text{because } 23 \div 6 = 3 \text{ with a remainder of } 5$$

◆**Example 4** Convert $\frac{45}{7}$ to a mixed number.

$$\frac{45}{7} = 6\frac{3}{7}. \qquad \text{because } 45 \div 7 = 6 \text{ with a remainder of } 3$$

Multiplication and Division of Mixed Numbers

To multiply or divide mixed numbers, first convert each mixed number to an improper fraction. Then multiply or divide as indicated. Be sure that all answers are left in the simplest (reduced) form.

◆**Example 5** Find the product. $3\frac{1}{2} \cdot 2\frac{2}{3}$

$$3\frac{1}{2} \cdot 2\frac{2}{3} = \frac{7}{2} \cdot \frac{8}{3} \qquad \text{Convert to improper fractions}$$

$$= \frac{7}{2} \cdot \frac{2 \cdot 2 \cdot 2}{3} \qquad \text{Factor into primes}$$

$$\boxed{\frac{28}{3} \text{ or } 9\frac{1}{3}}$$

Note: Both of the above are considered "simplest form". The important consideration is to determine if there is a common factor that can be reduced.

◆**Example 6** Find the quotient. $4\frac{1}{5} \div 3\frac{1}{2}$

$$4\frac{1}{5} \div 3\frac{1}{2} = \frac{21}{5} \div \frac{7}{2} \qquad \text{Convert to improper fractions}$$

$$= \frac{21}{5} \cdot \frac{2}{7} \qquad \text{Multiply first fraction by reciprocal of second}$$

$$= \frac{3 \cdot 7}{5} \cdot \frac{2}{7} \qquad \text{Factor into primes}$$

$$\boxed{\frac{6}{5} \text{ or } 1\frac{1}{5}}$$

Addition of Mixed Numbers

To add mixed numbers, add the whole number parts to the whole number parts and add the fractional parts to the fractional parts. (Similar to adding like terms)

◆ Example 7   Find the sum. $2\frac{1}{3} + 4\frac{1}{2}$

$$2\frac{1}{3} + 4\frac{1}{2} = 2\left(\frac{1}{3} \cdot \frac{2}{2}\right) + 4\left(\frac{1}{2} \cdot \frac{3}{3}\right) \quad \text{Multiply by 1 in the form of } \frac{2}{2} \text{ and } \frac{3}{3}$$

$$= 2\frac{2}{6} + 4\frac{3}{6}$$

$$\boxed{6\frac{5}{6}}$$

◆ Example 8   Find the sum. $4\frac{7}{10} + 2\frac{1}{8}$

$$4\frac{7}{10} + 2\frac{1}{8} = 4\left(\frac{7}{10} \cdot \frac{4}{4}\right) + 2\left(\frac{1}{8} \cdot \frac{5}{5}\right) \quad \text{Multiply by 1 in the form of } \frac{4}{4} \text{ and } \frac{5}{5}$$

$$= 4\frac{28}{40} + 2\frac{5}{40}$$

$$\boxed{6\frac{33}{40}}$$

◆ Example 9   Find the sum. $4\frac{7}{12} + 3\frac{5}{8}$

$$4\frac{7}{12} + 3\frac{5}{8} = 4\left(\frac{7}{12} \cdot \frac{2}{2}\right) + 3\left(\frac{5}{8} \cdot \frac{3}{3}\right) \quad \text{Multiply by 1 in the form of } \frac{2}{2} \text{ and } \frac{3}{3}$$

$$= 4\frac{14}{24} + 3\frac{15}{24} \quad \text{Convert to LCD in order to add fractions}$$

$$= 7\frac{29}{24} \quad \text{Add whole parts and add fractional parts}$$

$$= 7 + 1\frac{5}{24} \quad \text{Must have whole number and } \textbf{proper} \text{ fraction}$$

$$\boxed{8\frac{5}{24}}$$

Subtraction of Mixed Numbers

To subtract mixed numbers, subtract the whole number parts from the whole number parts and subtract the fractional parts from the fractional parts. As we shall see, this may involve a re-writing process called borrowing

◆ Example 10 Find the difference. $7\frac{1}{2} - 2\frac{2}{3}$

$$7\frac{1}{2} - 2\frac{2}{3} = 7\frac{3}{6} - 2\frac{4}{6}$$ Convert to LCD of 6

$$= 6 + 1\frac{3}{6} - 2\frac{4}{6}$$ We cannot subtract fractional part $\frac{4}{6}$ from $\frac{3}{6}$ so we will rewrite $7\frac{3}{6}$ to be $6 + 1\frac{3}{6}$

$$= 6\frac{9}{6} - 2\frac{4}{6}$$ Write $1\frac{3}{6}$ as $\frac{9}{6}$

$$\boxed{4\frac{5}{6}}$$

◆ Example 11 Find the difference. $8\frac{3}{5} - 5\frac{3}{4}$

$$8\frac{3}{5} - 5\frac{3}{4} = 8\left(\frac{3}{5} \cdot \frac{4}{4}\right) - 5\left(\frac{3}{4} \cdot \frac{5}{5}\right)$$

$$= 8\frac{12}{20} - 5\frac{15}{20}$$

$$= 7 + 1\frac{12}{20} - 5\frac{15}{20}$$

$$= 7\frac{32}{20} - 5\frac{15}{20}$$

$$\boxed{2\frac{17}{20}}$$

Combining Operations

When more than one operation appears in an expression, we must remember to follow the order of operations established in Section 1.1.

1. Parenthesis (any grouping).

2. Exponents.

3. Multiplication or division, whichever comes first working left to right.

4. Addition or subtraction, whichever comes first working left to right.

◆**Example 12**   Perform the indicated operation. $3\frac{1}{2} \cdot 1\frac{3}{5} + 3\frac{7}{10}$

$3\frac{1}{2} \cdot 1\frac{3}{5} + 3\frac{7}{10} = \frac{7}{2} \cdot \frac{8}{5} + 3\frac{7}{10}$ Multiply before add, convert to improper fractions

$= \frac{7}{2} \cdot \frac{2 \cdot 2 \cdot 2}{5} + 3\frac{7}{10}$ Factor into primes

$= \frac{28}{5} + 3\frac{7}{10}$ Cancel, multiply

$= 5\frac{3}{5} + 3\frac{7}{10}$ Change to mixed number

$= 5\left(\frac{3}{5} \cdot \frac{2}{2}\right) + 3\frac{7}{10}$ Convert to LCD by multiplying by 1 in the form of $\frac{2}{2}$

$= 5\frac{6}{10} + 3\frac{7}{10}$

$= 8\frac{13}{10}$

$= 8 + 1\frac{3}{10}$

$\boxed{9\frac{3}{10}}$

Solving Word Problems Having Mixed Numbers

Word problems involving mixed numbers will follow the same pattern as the word problems with integers discussed in Section 1.1. We will **organize** information given in the problem, look for **patterns** within the information, make **decisions** about how we will use the information, use **skills** to simplify expressions and answer the question with a **sentence**.

◆**Example 13**   Solve the following word problem.

Arturo made a 12 pound salad for the math picnic. He used $3\frac{1}{2}$ pounds of tomatoes, $4\frac{1}{3}$ pounds of onions (math students like onions), $2\frac{3}{4}$ pounds of cucumbers and the remainder was lettuce. How much lettuce did he use?

Organize information: $3\frac{1}{2} + 4\frac{1}{3} + 2\frac{3}{4}$ = Total of known quantities

12 = Total number of pounds in the salad

Pattern: Remainder is obtained in a subtraction problem.

Decision: Total amount of salad − total amount of known quantities = remainder

$$12 - \left(3\frac{1}{2} + 4\frac{1}{3} + 2\frac{3}{4}\right) = \text{remainder}$$

Skill:
$$12 - \left(3\frac{6}{12} + 2\frac{9}{12} + 4\frac{4}{12}\right) = \text{remainder}$$

$$12 - \left(9\frac{19}{12}\right) = \text{remainder}$$

$$12 - \left(10\frac{7}{12}\right) = \text{remainder}$$

$$11\frac{12}{12} - \left(10\frac{7}{12}\right) = \text{remainder}$$

$$1\frac{5}{12} = \text{remainder}$$

Sentence: Arturo used $1\frac{5}{12}$ pounds of lettuce in the salad.

◆**Example 14** Solve the following word problem.

Omar spent $18 making dessert for 10 people. He used 6 ounces of pecans, $11\frac{2}{3}$ ounces of whipping cream and 24 ounces of strawberries. He was careful to see that each portion was approximately of equal size. How much whipping cream did the average dessert contain?

Organize information: $11\frac{2}{3}$ ounces = amount of whipping cream

10 = number of desserts made

Pattern: To find average, divide the amount of cream by the number of desserts.

Decision: $11\frac{2}{3} \div 10$

Skill: $11\frac{2}{3} \div 10 = \frac{35}{3} \cdot \frac{1}{10}$

$$= \frac{7}{6} \text{ or } 1\frac{1}{6}$$

Sentence: Each dessert used $1\frac{1}{6}$ ounces of whipping cream.

Express each of the following fractions as a mixed fraction.

1) $\frac{8}{3} =$

2) $\frac{12}{5} =$

3) $\frac{19}{4} =$

Express each of the following mixed fractions as an improper fraction.

4) $3\frac{2}{5} =$

5) $6\frac{1}{3} =$

6) $4\frac{3}{4} =$

Perform the indicated operation.

7) $7\frac{2}{3} + 5\frac{5}{6}$

8) $2\frac{3}{8} + 4\frac{3}{4}$

9) $8\frac{7}{12} - 3\frac{1}{4}$

10) $2\frac{1}{3} \cdot 1\frac{3}{14}$

11) $3\frac{1}{4} \div \frac{3}{8}$

12) $2\frac{3}{4} \cdot 2\frac{2}{3}$

13) $3\frac{5}{6} + 2\frac{2}{3}$

14) $7\frac{3}{4} - 2\frac{1}{2}$

15) $3\frac{1}{2} \cdot 2\frac{2}{5}$

16) $8\frac{1}{2} - 3\frac{5}{6}$

17) $8\frac{1}{3} \div \frac{5}{18}$

18) $6 - 3\frac{2}{7}$

4.4 Problem Set

Perform the indicated operations.

19) $3\frac{1}{2} \cdot \frac{3}{8} \div \frac{5}{16}$	20) $4\frac{1}{4} + 5\frac{1}{3} \div \frac{8}{9}$	21) $2\frac{2}{5} \cdot 3\frac{3}{4} + 3\frac{5}{8}$
22) $5\frac{2}{3} \div 4\frac{2}{9} + 3\frac{1}{2}$	23) $3\frac{1}{5}\left[1\frac{1}{5} + 2\frac{1}{3}\right]$	24) $1\frac{3}{16} + \frac{3}{8} \cdot \frac{5}{9}$
25) $\left[\frac{5}{4}\right]^2 + 3\frac{1}{2} \cdot \frac{19}{56}$	26) $3\frac{1}{5} + 1\frac{1}{6} \cdot \frac{3}{5}$	27) $2\frac{1}{4} \cdot 2\frac{2}{5} - 1\frac{1}{3} \div \frac{5}{12}$

Solve the following word problems.

28) For Mother's Day, Kent baked a cake which called for $2\frac{1}{4}$ cups of flour and $1\frac{1}{2}$ cups of sugar. Because everyone seemed to be on a diet, he decided to use only $\frac{2}{3}$ of the recommended ingredients. How much flour should he use?

29) Mrs. Rodriguez was $24\frac{1}{2}$ years old when she had her first child, Elmo, who weighed a whopping 9 pounds. Little Sarah arrived 30 months later weighing in at $6\frac{2}{3}$ pounds. How old was Mrs. Rodriguez when Sarah arrived?

30) Fannie Farkel needed $12\frac{4}{5}$ yards of material to make four table cloths. How much material did she use on each table cloth?

31) Valerie paid $48 for ten pounds of New Mexico pecans. She used $1\frac{1}{4}$ pounds to make chocolate chip cookies, $\frac{7}{8}$ of a pound in a salad, and ate one-half pound while she was cooking. How many pounds of pecans does she have remaining?

32) Alex uses $2\frac{3}{8}$ feet of ribbon on each Christmas wreath that he makes. If he made 28 wreaths, how much ribbon did he use?

33) Laurel and Hardy purchased 32 feet of lumber for $98 to build a special piano moving platform. They used four boards measuring $2\frac{1}{2}$ feet long and eight boards measuring $1\frac{3}{4}$ feet long. If they sold the remaining boards to Tarzan at $3 per foot, how much did Tarzan pay for the boards?

34) Michael Barisnakov bought 120 yards of a material to make special dance outfits for his new tour. Each of the 24 outfits required $3\frac{2}{3}$ yards of material. If Delta Burke bought all of the remaining material at $14 per yard, how much did it cost her?

35) Steve Martin rode his bike for two hours and 20 minutes and averaged 13 MPH. How far did he ride? (Use the formula d = r • t)

4.5 Translations Involving Fractions

In Section 1.4, we translated words into symbols using integers. We will now translate words into symbols using fractions.

◆**Example 1**   Find the results when the quotient of $8\frac{3}{4}$ and $3\frac{1}{8}$ is decreased by $1\frac{3}{5}$

$$8\frac{3}{4} \div 3\frac{1}{8} - 1\frac{3}{5} \qquad \text{Translation}$$

$$= \frac{35}{4} \div \frac{25}{8} - 1\frac{3}{5}$$

$$= \frac{35}{4} \cdot \frac{8}{25} - 1\frac{3}{5}$$

$$= \frac{14}{5} - 1\frac{3}{5}$$

$$= 2\frac{4}{5} - 1\frac{3}{5}$$

$$\boxed{1\frac{1}{5} \text{ is the result.}}$$

◆**Example 2**   Add $2\frac{1}{4}$ to the product of $3\frac{1}{2}$ and $1\frac{3}{5}$.

$$2\frac{1}{4} + \left(3\frac{1}{2}\right)\left(1\frac{3}{5}\right) \qquad \text{Translation}$$

$$= 2\frac{1}{4} + \frac{7}{2} \cdot \frac{8}{5}$$

$$= 2\frac{1}{4} + \frac{28}{5}$$

$$= 2\frac{1}{4} + 5\frac{3}{5}$$

$$= 2\left(\frac{1}{4} \cdot \frac{5}{5}\right) + 5\left(\frac{3}{5} \cdot \frac{4}{4}\right)$$

$$= 2\frac{5}{20} + 5\frac{12}{20}$$

$$\boxed{7\frac{17}{20} \text{ is the result.}}$$

Chapter 4: Fractions

◆**Example 3** Multiply $2\frac{2}{3}$ by the sum of $2\frac{3}{4}$ and $3\frac{1}{2}$

$$2\frac{2}{3}\left(2\frac{3}{4}+3\frac{1}{2}\right) \quad \text{Translation}$$

$$=2\frac{2}{3}\left[2\frac{3}{4}+3\left(\frac{1}{2}\cdot\frac{2}{2}\right)\right]$$

$$=2\frac{2}{3}\left[2\frac{3}{4}+3\frac{2}{4}\right]$$

$$=2\frac{2}{3}\left[5\frac{5}{4}\right]$$

$$=\frac{8}{3}\cdot\frac{25}{4}$$

$$=\frac{50}{3} \quad \text{or} \quad 16\frac{2}{3}$$

The result is $16\frac{2}{3}$.

◆**Example 4** Subtract $2\frac{7}{8}$ from the square of $2\frac{1}{2}$

$$\left(2\frac{1}{2}\right)^2-2\frac{7}{8} \quad \text{Translation}$$

$$=\left(\frac{5}{2}\right)^2-2\frac{7}{8}$$

$$=\frac{25}{4}-2\frac{7}{8}$$

$$=6\frac{1}{4}-2\frac{7}{8}$$

$$=6\left(\frac{1}{4}\cdot\frac{2}{2}\right)-2\frac{7}{8}$$

$$=6\frac{2}{8}-2\frac{7}{8}$$

$$=5\frac{10}{8}-2\frac{7}{8}$$

The result is $3\frac{7}{8}$.

Section 4.5: Translations Using Fractions

Helpful Hint 10 : To translate a subtraction problem using the word "from", be sure to place the "from number" first. For example, the expression "subtract two from eight" translates as 8–2. The number 8 in this case is referred to as the "from number".

◆**Example 5** Add $3\frac{1}{8}$ to the square of the sum of $\frac{2}{3}$ and $\frac{5}{6}$

$$3\frac{1}{8} + \left(\frac{2}{3} + \frac{5}{6}\right)^2 \qquad \text{Translation}$$

$$= 3\frac{1}{8} + \left(\frac{2}{3} \cdot \frac{2}{2} + \frac{5}{6}\right)^2$$

$$= 3\frac{1}{8} + \left(\frac{4}{6} + \frac{5}{6}\right)^2$$

$$= 3\frac{1}{8} + \left(\frac{9}{6}\right)^2$$

$$= 3\frac{1}{8} + \left(\frac{3}{2}\right)^2$$

$$= 3\frac{1}{8} + \frac{9}{4}$$

$$= 3\frac{1}{8} + 2\frac{1}{4}$$

$$= 3\frac{1}{8} + 2\frac{2}{8}$$

$$\boxed{\text{The result is } 5\frac{3}{8}.}$$

Name:_____

Perform the indicated operations.

1) $6\frac{2}{3} + 5\frac{1}{2} - 2\frac{5}{6}$

2) $1\frac{4}{7} \cdot 4\frac{2}{3} + 2\frac{1}{3}$

3) $4\frac{1}{6} \div \frac{5}{11} - 1\frac{1}{6}$

4) $1\frac{2}{7} + 2\frac{3}{5} \cdot 1\frac{3}{7}$

5) $8\frac{1}{4} - 5\frac{5}{6} + 2\frac{2}{3}$

6) $5\frac{1}{3} - 2\frac{3}{4} + 4\frac{7}{12}$

7) $4\frac{3}{5} \cdot 1\frac{3}{7} \cdot 5\frac{1}{4}$

8) $8\frac{5}{8} + 3\frac{5}{6} - 6\frac{3}{4}$

9) $\left[5\frac{1}{2} + 3\frac{2}{3}\right] \div 1\frac{2}{9}$

Perform the indicated operations.

10) $2\frac{2}{3} \div 1\frac{5}{6} \cdot 4\frac{2}{5}$

11) $24 \div \frac{3}{7} + 2\frac{3}{4}$

12) $\left[\frac{5}{2}\right]^2 + 2\frac{2}{5} \cdot \left[2\frac{1}{4} + 1\frac{5}{6}\right]$

13) $1\frac{7}{12} + 8\frac{3}{4} - 4\frac{5}{6}$

14) $2\frac{1}{2} + \frac{9}{14} \cdot \left[1\frac{5}{8} + 2\frac{3}{4}\right]$

15) $\left[\frac{2}{3}\right]^2 \cdot 1\frac{1}{5} + \frac{5}{8}\left[\frac{4}{3}\right]^2$

Solve the following problems.

16) Rocco paid exactly $5 for the ingredients for his famous salad. He used $2\frac{1}{4}$ pounds of tomatoes, $1\frac{2}{3}$ pounds of cucumbers, $4\frac{3}{4}$ pounds of lettuce and $\frac{2}{3}$ pounds of garlic. By 8 PM he and his six guests had eaten the entire salad. On the average how much did each person eat?

17) Fannie Farkel entered a cherry picking contest. The first minute she picked $1\frac{3}{4}$ pounds, the second minute she picked $2\frac{3}{8}$ pounds, and the third minute she picked a whopping $3\frac{1}{2}$ pounds. Tarzan won the contest with a record setting $9\frac{3}{4}$ pounds. By how many pounds did Tarzan beat Fannie?

Translate and simplify.

18) Find the result when the square of $\frac{3}{5}$ is added to the product of $\frac{4}{5}$ and $\frac{2}{3}$.

19) Multiply $2\frac{2}{3}$ by the sum of $\frac{5}{8}$ and $\frac{3}{4}$.

20) Multiply $3\frac{1}{3}$ by the difference of $5\frac{2}{3}$ and $2\frac{1}{6}$.

21) Add $4\frac{3}{8}$ to the product of $1\frac{1}{5}$ and $3\frac{3}{4}$.

22) Subtract $2\frac{3}{4}$ from the product of $1\frac{4}{5}$ and $6\frac{2}{3}$.

23) Subtract $1\frac{1}{4}$ from the product of $1\frac{1}{6}$ and $1\frac{7}{8}$.

24) Add the square of $\frac{3}{4}$ to the product of $\frac{5}{8}$ and $2\frac{3}{5}$.

25) Add $1\frac{7}{8}$ to the square of the difference of $5\frac{1}{2}$ and $3\frac{3}{4}$.

4.6 Evaluating Algebraic Expressions Using Fractions

In this section we will continue to review operations on fractions and add to our skill base by evaluating expressions that contain fractions.

Evaluating an algebraic expression

Throughout Chapter 1, we evaluated algebraic expressions using integers. We will now use the same organization as we extend this skill to include fractions. The key to evaluating an algebraic expression is to recopy the problem with all indicated operations in place but using empty parenthesis every time a variable appears. The next step is to fill-in the empty parenthesis with the indicated values of the variables.

◆**Example 1** Evaluate $a[x - n]$ if $\quad a = \dfrac{7}{8} \qquad x = \dfrac{5}{6} \qquad n = \dfrac{1}{2}$

$\left(\dfrac{7}{8}\right)\left[\left(\dfrac{5}{6}\right) - \left(\dfrac{1}{2}\right)\right]$ Empty parenthesis, fill in

$= \dfrac{7}{8}\left[\dfrac{5}{6} - \dfrac{1}{2} \cdot \dfrac{3}{3}\right]$

$= \dfrac{7}{8}\left[\dfrac{5}{6} - \dfrac{3}{6}\right]$

$= \dfrac{7}{8}\left[\dfrac{2}{6}\right]$

$= \dfrac{7}{8} \cdot \dfrac{1}{3}$

$\boxed{= \dfrac{7}{24}}$

◆**Example 2** Evaluate $bc + w$ if $\quad b = 3\dfrac{1}{2} \qquad c = \dfrac{5}{7} \qquad w = 3\dfrac{1}{4}$

$= \dfrac{7}{2} \cdot \dfrac{5}{7} + 3\dfrac{1}{4}$

$= \dfrac{5}{2} + 3\dfrac{1}{4}$

$= 2\dfrac{1}{2} + 3\dfrac{1}{4}$

$= 2\left(\dfrac{1}{2} \cdot \dfrac{2}{2}\right) + 3\dfrac{1}{4}$

$= 2\dfrac{2}{4} + 3\dfrac{1}{4}$

$\boxed{= 5\dfrac{3}{4}}$

Perform the indicated operations.

1) $5\frac{3}{4} + 2\frac{5}{6} - 3\frac{2}{3}$

2) $1\frac{3}{5} \cdot 3\frac{1}{3} + \frac{7}{8} \cdot 3\frac{1}{5}$

3) $2\frac{2}{5}\left[5\frac{1}{3} - 2\frac{1}{2}\right]$

4) $6\frac{1}{4} \div 1\frac{7}{8} + 2\frac{1}{7} \cdot \frac{7}{10}$

5) $2\frac{1}{7}\left[8\frac{2}{5} - 3\frac{1}{2}\right]$

6) $1\frac{1}{4}\left[5\frac{3}{4} + 3\frac{5}{6}\right]$

7) $5\frac{1}{3}\left[12\frac{7}{8} - 5\frac{3}{4}\right]$

8) $2\frac{7}{9}\left[15\frac{2}{5} - 12\frac{7}{10}\right]$

9) $3\frac{2}{3} + 3\frac{1}{3}\left[6 - 4\frac{2}{5}\right]$

4.6 Problem Set

Evaluate each expression using $a = \dfrac{3}{4}$ $b = \dfrac{2}{3}$ $n = \dfrac{5}{8}$ $x = \dfrac{5}{2}$ $w = \dfrac{7}{10}$

10) ab + nw	11) b[x − w]
12) ab − ax	13) n + x[a − n]

Solve each of the following problems.

14) Jung Han ran an average of 8 miles per hour for 2 hours and 40 minutes. How far did he run? (Use the distance formula d = r • t)	15) The Ferrari family bought 10 pounds of peaches. Gino ate $\dfrac{2}{5}$ of the total, Maria ate $\dfrac{1}{8}$ of the total and Salvatore ate the rest. How many pounds did Salvatore eat?

16) Nathan can ride his bike an average of 14 MPH. Gertraude can hop on her pogo stick an average of 2 MPH. If they started at the same spot at 2 PM and went in the same direction until 3:15 PM, how far ahead would Nathan be?

17) Arnold weighs $2\frac{1}{2}$ times the weight of his 24 pound sister Sue. Their father Fred weighs $2\frac{1}{4}$ times their combined weight. How much does their father weigh?

18) Add $3\frac{5}{6}$ to the product of $2\frac{2}{3}$ and $1\frac{3}{4}$.

19) Julie's boat can go 12 MPH in still water. If the current flows at $1\frac{3}{4}$ MPH in Shady River and she went for an hour and 45 minutes downstream, how far did she travel? (The current will add to the speed of the boat going downstream.)

20) Alberto can paddle his canoe a solid 6 MPH in still water. One sunny afternoon he decided to paddle upstream in a river that had a current of $1\frac{1}{2}$ MPH. If he started at 1 PM, how far had he traveled by 2:40 PM? (The current will decrease the speed of the boat going upstream.)

21) Rocco made a fruit salad. He used $3\frac{1}{4}$ pounds of oranges, $3\frac{1}{3}$ pounds of peaches and $2\frac{1}{2}$ pounds of bananas. How many pounds of strawberries did he use to complete the 12 pound salad?

22. Make up and solve a word problem using fractions.

4.7 Addition or Subtraction with Large Denominators

In a problem of addition or subtraction of fractions, the LCD is often apparent. If the denominators are large, finding the LCD may not come from a quick glance. This section develops an organized approach to use, if the LCD is not quickly observed.

A Strategy to Apply When Denominators are Large

In Section 4.3, we developed a process for finding the LCM for a set of numbers. We will now apply this process to develop an organization to use on addition or subtraction of fractions with large denominators. The following steps will be used in our organization:

1. Factor each of the denominators into prime factors.
2. The LCD will use each factor the most number of times it is used in any one of the denominators.
3. Change each term into a term having the LCD.
4. Add or subtract in the usual manner.

◆**Example 1** Perform the indicated operations. $3\frac{5}{8} + 2\frac{7}{12} - 4\frac{3}{20}$

$$\left.\begin{array}{l}8 = 2 \cdot 2 \cdot 2 \\ 12 = 2 \cdot 2 \cdot 3 \\ 20 = 2 \cdot 2 \cdot 5\end{array}\right\}$$ Factor each denominator into primes

$\text{LCD} = 2 \cdot 2 \cdot 2 \cdot 3 \cdot 5 = 120$ Each factor is used the most

$= 3\left(\frac{5}{8} \cdot \frac{15}{15}\right) + 2\left(\frac{7}{12} \cdot \frac{10}{10}\right) - 4\left(\frac{3}{20} \cdot \frac{6}{6}\right)$ Change to LCD

$= 3\left(\frac{75}{120}\right) + 2\left(\frac{70}{120}\right) - 4\left(\frac{18}{120}\right)$ Simplify

$= 5\frac{145}{120} - 4\frac{18}{120}$

$= 1\frac{127}{120}$

$$\boxed{2\frac{7}{120}}$$

Name:_____   4.7 Problem Set

Use the model of problem #1 to organize each of the remaining problems.

1) $5\frac{3}{4} + 2\frac{7}{12} + 3\frac{1}{18}$ Factor denominators: $\quad 4 = (2)(2)$ $\quad 12 = (2)(2)(3)$ $\quad 18 = (2)(3)(3)$ $\quad LCD = (2)(2)(3)(3) = 36$ Note: Each factor is used "the most" that it is used in any one of the terms. Change to LCD: $5\frac{\ }{36} + 2\frac{\ }{36} + 3\frac{\ }{36}$	2) $4\frac{3}{8} + 2\frac{4}{15} + 5\frac{1}{6}$ $8 =$ $15 =$ $6 =$ $LCD =$ $4\frac{\ }{\ } + 2\frac{\ }{\ } + 5\frac{\ }{\ }$
3) $3\frac{5}{14} + 2\frac{5}{6} + 5\frac{8}{21}$ $14 =$ $6 =$ $21 =$ $LCD =$	4) $6\frac{7}{10} + 3\frac{3}{8} + 2\frac{3}{20}$
5) $8\frac{8}{21} - 2\frac{2}{35}$	6) $9\frac{7}{24} - 4\frac{9}{20}$

7) $8\frac{3}{8} + 5\frac{8}{15} - 7\frac{9}{20}$

8) $2\frac{5}{33} + 7\frac{5}{22} - 3\frac{2}{3}$

9) $4\frac{3}{14} + 2\frac{9}{35} - 3\frac{3}{20}$

10) $9\frac{3}{4} - 5\frac{3}{28} - 2\frac{8}{21}$

11) $8\frac{5}{8} - 5\frac{7}{24} + 2\frac{4}{9}$

12) $4\frac{5}{6} - 3\frac{3}{8} + 4\frac{7}{10} - 2\frac{11}{15}$

4.8 Word Problems Having Fractions with Large Denominators

In this section, we will continue to practice the organization required to simplify addition and subtraction of fractions having large denominators. We will also apply this organization to the word problems that have large denominators.

Organization for Word Problems

The organization for solving word problems with fractions having large denominators is the same as the organization for word problems having smaller denominators.

1. **Organize** the information.
2. Look for **patterns** in the information given.
3. Make **decisions** about the patterns within the information.
4. Use **skills** to simplify or solve.
5. Write a **sentence** that answers the question or demand of the problem.

◆ **Example 1** Use the organization listed above to solve the following problem:

Roosevelt Greer at age 28, arrived at football camp weighing 278 pounds. During the first week of camp he gained $4\frac{9}{20}$ pounds. Wanting to trim down a bit, Roosevelt then lost $6\frac{5}{12}$ pounds the second week and lost another $8\frac{7}{15}$ pounds during the third week. How much did Roosevelt weigh at the end of the third week?

Organize information:

278 pounds	Starting weight
$4\frac{9}{20}$ pounds	Number of pounds **gained** during first week
$6\frac{5}{12}$ pounds	Number of pounds **lost** during second week
$8\frac{7}{15}$ pounds	Number of pounds **lost** during second week

Patterns: To **gain** weight is to **add**; to **lose** weight is to **subtract**

Decision: Total beginning weight added to pounds gained minus total pounds lost

$$\text{New weight} = 278 + 4\frac{9}{20} - 6\frac{5}{12} - 8\frac{7}{15}$$

Use **skill** to simplify:

$$\text{New weight} = 278 + 4\frac{9}{20} - 6\frac{5}{12} - 8\frac{7}{15}$$

$$\left.\begin{array}{l}20 = 2 \cdot 2 \cdot 5 \\ 12 = 2 \cdot 2 \cdot 3 \\ 15 = 3 \cdot 5\end{array}\right\} \quad \text{Factor into primes}$$

$$\text{LCD} = 60$$

$$\text{New weight} = 278 + 4\left(\frac{9}{20} \cdot \frac{3}{3}\right) - 6\left(\frac{5}{12} \cdot \frac{5}{5}\right) - 8\left(\frac{7}{15} \cdot \frac{4}{4}\right) \quad \text{Change to LCD}$$

$$= 278 + 4\frac{27}{60} - 6\frac{25}{60} - 8\frac{28}{60} \quad \text{Simplify}$$

$$= 282\frac{27}{60} - 6\frac{25}{60} - 8\frac{28}{60} \quad \text{Addition before subtraction (left to right)}$$

$$= 276\frac{2}{60} - 8\frac{28}{60} \quad \text{Subtraction}$$

$$= 275 + 1\frac{2}{60} - 8\frac{28}{60} \quad \text{Rewrite } 276 = 275 + 1$$

$$= 275\frac{62}{60} - 8\frac{28}{60} \quad \text{Rewrite } 1\frac{2}{60} \text{ as } \frac{62}{60}$$

$$= 267\frac{34}{60} \quad \text{Subtract}$$

$$= 267\boxed{\frac{17}{30}} \quad \text{Reduce}$$

Roosevelt weighed $267\frac{17}{30}$ pounds at the end of the third week

---

Perform the indicated operations.

1) $5\frac{7}{12} + 2\frac{3}{10} + 3\frac{4}{15}$

2) $2\frac{5}{8} + 3\frac{7}{20} + 4\frac{4}{15}$

3) $5\frac{3}{20} + 2\frac{7}{12} - 4\frac{5}{18}$

4) $6\frac{7}{10} + 2\frac{5}{12} - 5\frac{8}{15}$

5) $8\frac{4}{15} - 3\frac{7}{20} + 2\frac{5}{6}$

6) $4\frac{7}{8} - 5\frac{2}{27} + 3\frac{5}{12}$

7) For a special trail mix, Wilbert mixed $2\frac{3}{8}$ pounds of almonds, $3\frac{7}{12}$ pounds of dried apricots and $4\frac{5}{6}$ pounds of dried apples. During his week-long hike, he ate the whole mix. On the average, how much did he eat each day?

8) For a big jungle feast, Tarzan mixed $4\frac{3}{8}$ pounds of tomatoes, $2\frac{5}{6}$ pounds of cucumbers and $7\frac{3}{10}$ pounds of lettuce. How much garlic should he add so that he will have exactly 15 pounds of salad?

9) Find the result when $6\frac{5}{12}$ is diminished by the product of $1\frac{7}{8}$ and $1\frac{5}{27}$.

10) Find the result when the sum of $1\frac{5}{8}$ and $2\frac{1}{6}$ is divided by $\frac{5}{48}$.

4.9 Chapter Review

In this chapter, you learned to:

1. Operate on fractions.
2. Solve word problems using fractions.

Multiplication of Common Fractions [4.2]

To multiply common fractions:

1. Factor the numerator and denominator into primes.
2. Cancel factors that are common to the numerators and denominators.
3. The product of the remaining numerators is the numerator of the answer.
4. The product of the remaining denominators is the denominator of the answer.

◆**Example 1** Find the product: $\dfrac{7}{12} \cdot \dfrac{9}{14}$

$$= \dfrac{7}{2 \cdot 2 \cdot 3} \cdot \dfrac{3 \cdot 3}{2 \cdot 7}$$

$$\boxed{= \dfrac{3}{8}}$$

Division of Common Fractions [4.2]

To divide common fractions:

1. Multiply the first fraction by the reciprocal of the second fraction.
2. Use the procedures outlined in the multiplication of common fraction process.

◆**Example 2** Find the quotient: $\dfrac{9}{2} \div \dfrac{15}{4}$

$$= \dfrac{9}{2} \cdot \dfrac{4}{15}$$

$$= \dfrac{3 \cdot 3}{2} \cdot \dfrac{2 \cdot 2}{3 \cdot 5}$$

$$\boxed{= \dfrac{6}{5}}$$

Addition of Common Fractions   [4.2]

To add common fractions:

1. If the denominators are the same, proceed to step 2. If the denominator are different, change each fraction to the LCD.
2. The sum of the numerators is the numerator of the answer.
3. The common denominator is the denominator of the answer.
4. Reduce to lowest term if possible.

◆ **Example 3** Find the sum: $\dfrac{5}{6} + \dfrac{3}{4}$

$$= \dfrac{5}{6} \cdot \dfrac{2}{2} + \dfrac{3}{4} \cdot \dfrac{3}{3}$$
$$= \dfrac{10}{12} + \dfrac{9}{12}$$

$$\boxed{= \dfrac{19}{12}}$$

Subtraction of Common Fractions   [4.2]

To subtract common fractions:

1. If the denominators are the same, proceed to step 2. If the denominator are different, change each fraction to the LCD.
2. The difference of the numerators is the numerator of the answer.
3. The common denominator is the denominator of the answer.
4. Reduce to lowest term if possible.

◆ **Example 4** Find the difference: $\dfrac{7}{8} - \dfrac{5}{6}$

$$= \dfrac{7}{8} \cdot \dfrac{3}{3} - \dfrac{5}{6} \cdot \dfrac{4}{4}$$
$$= \dfrac{21}{24} - \dfrac{20}{24}$$

$$\boxed{= \dfrac{1}{24}}$$

Multiplication or Division of Mixed Numbers   [4.4]

To multiply mixed numbers:

1. Convert each mixed number to an improper fraction.
2. Multiply or divide using the procedure for common fractions.

◆**Example 5** Find the quotient: $3\frac{3}{4} \div 1\frac{1}{8}$

$$= \frac{15}{4} \div \frac{9}{8}$$ Convert to improper fractions

$$= \frac{15}{4} \cdot \frac{8}{9}$$ Multiply first fraction by reciprocal of second fraction.

$$= \frac{5 \cdot 3}{2 \cdot 2} \cdot \frac{2 \cdot 2 \cdot 2}{3 \cdot 3}$$ Factor into primes, cancel.

$$\boxed{= \frac{10}{3} \text{ or } 3\frac{1}{3}}$$

Addition of Mixed Numbers   [4.4]

To add mixed numbers:

1. If the denominators are the same, proceed to step 2. If the denominator are different, change each fractional part to the LCD.
2. Add whole number parts to whole number parts, and fractional parts to fractional parts.
3. Reduce fractional parts if possible.
4. If fractional part is improper, change to mixed number and add to the whole number.

◆**Example 6** Find the sum. $5\frac{3}{4} + 1\frac{5}{6}$

$$= 5\left(\frac{3}{4} \cdot \frac{3}{3}\right) + 1\left(\frac{5}{6} \cdot \frac{2}{2}\right)$$ Change to LCD.

$$= 5\frac{9}{12} + 1\frac{10}{12}$$

$$= 6\frac{19}{12}$$ Add whole parts, add fractional parts

$$= 6 + 1\frac{7}{12}$$ Change fractional part to common fraction

$$\boxed{= 7\frac{7}{12}}$$

Subtraction of Mixed Numbers   [4.4]

To subtract mixed numbers:

1. If the denominators are the same, proceed to step 2. If the denominator are different, change each fractional part to the LCD.

2. If the second fractional part will subtract from the first fractional part proceed to step 3. If the second fractional part is larger than the first fractional part, then the first fractional part must be rewritten using the technique of borrowing.

3. Subtract whole number parts from whole number parts, and fractional parts from fractional parts.

4. Reduce fractional parts if possible.

5. If fractional part is an improper fraction, change to mixed number form and add to the whole number.

◆**Example 7**   Find the difference: $8\frac{3}{8} - 2\frac{3}{4}$

$$= 8\frac{3}{8} - 2\left(\frac{3}{4} \cdot \frac{2}{2}\right) \quad \text{Change to LCD.}$$

$$= 8\frac{3}{8} - 2\frac{6}{8}$$

$$= 7 + 1\frac{3}{8} - 2\frac{6}{8} \quad \text{Rewrite so subtraction is possible}$$

$$= 7\frac{11}{8} - 2\frac{6}{8}$$

$$\boxed{= 5\frac{5}{8}}$$

Addition or Subtraction of Fractions with Large Denominators   [4.7]

When denominators are so large that the LCD is not quickly apparent, use the following procedure:

1. Factor each of the denominators into primes.

2. The LCD will use each factor the most number of times its is used in any one of the denominators.

3. Change each term into a term having the LCD.

4. Add or subtract in the usual manner.

◆**Example 8**   Perform the indicated operation:

$$5\frac{3}{8} + 4\frac{7}{15} - 2\frac{5}{12}$$

$$8 = 2 \cdot 2 \cdot 2$$
$$15 = 3 \cdot 5$$
$$12 = 2 \cdot 2 \cdot 3$$
$$LCD = 2 \cdot 2 \cdot 2 \cdot 3 \cdot 5 = 120$$

$$5\left(\frac{3}{8} \cdot \frac{15}{15}\right) + 4\left(\frac{7}{15} \cdot \frac{8}{8}\right) - 2\left(\frac{5}{12} \cdot \frac{10}{10}\right)$$

$$= 5\frac{45}{120} + 4\frac{56}{120} - 2\frac{50}{120}$$

$$= 9\frac{101}{120} - 2\frac{50}{120}$$

$$= 7\frac{51}{120}$$

$$\boxed{= 7\frac{17}{40}}$$

Word Problems Using Fractions   [4.4, 4.5]

To organize the solution for word problems:

1. **Organize** the information.
2. Look for **patterns**.
3. Use **skills** to simplify or solve.
4. Write a **sentence** that answers the question or demand of the problem.

◆**Example 9**   Use the word problem organization to solve the following problem.

Juanita can paddle her canoe $6\frac{1}{2}$ MPH in still water. One lovely Saturday morning after doing an invigorating $2\frac{3}{4}$ hours of mathematics homework, she paddled her canoe $1\frac{1}{5}$ hours down the American River where the current was flowing at a steady $1\frac{1}{3}$ MPH. How far did she travel on the river?

Organize information:

$6\frac{1}{2}$ MPH   rate in still water.

$1\frac{1}{3}$ MPH   rate of the current.

$1\frac{1}{5}$ hours   time traveled on the river.

While it is interesting to know that Juanita spent $2\frac{3}{4}$ hours doing her math homework, it is of no value to us in this problem.

Patterns:

Rate downstream is the rate in still water added to rate of the current.

Distance is rate multiplied by time (d = r • t).

Decision:

The question asks for distance, so distance = rate • time

$$\text{distance} = \left(6\frac{1}{2} + 1\frac{1}{3}\right) \cdot \left(1\frac{1}{5}\right)$$

rate downstream • time

Skill:

$$\left(6\frac{1}{2} + 1\frac{1}{3}\right) \cdot \left(1\frac{1}{5}\right)$$

$$= \left(6\frac{3}{6} + 1\frac{2}{6}\right) \cdot \left(1\frac{1}{5}\right)$$

$$= \left(7\frac{5}{6}\right) \cdot \left(1\frac{1}{5}\right)$$

$$= \frac{47}{6} \cdot \frac{6}{5}$$

$$= \frac{47}{5} \text{ or } 9\frac{2}{5}$$

Write a **sentence:**

> Juanita traveled $9\frac{2}{5}$ miles down the American River.

Name:_____

Perform the indicated operations.

1) $2\frac{3}{8} + 4\frac{1}{2} + 3\frac{3}{4}$	2) $6\frac{5}{8} + 2\frac{3}{4} - 4\frac{1}{2}$	3) $\left(2\frac{2}{5}\right) \cdot \left(3\frac{3}{4}\right) + 3\frac{2}{3}$
4) $3\frac{3}{5} \div 1\frac{5}{7} + 3\frac{3}{10}$	5) $4\frac{4}{5} \div \frac{8}{13} + 1\frac{3}{5}$	6) $1\frac{5}{7} \cdot \left(3\frac{1}{2} + 2\frac{2}{3}\right)$
7) $\frac{9}{20} \cdot \left(2\frac{1}{6} + 1\frac{2}{3}\right)$	8) $\left(5\frac{7}{8} - 2\frac{1}{4}\right) \div \frac{5}{24}$	9) $2\frac{7}{20} + 3\frac{4}{15} + 1\frac{7}{18}$

4.9 Chapter Review Test   Name:_____

Solve the following equations.

10) $7 + 5(x - 3) = 2(x + 14)$

11) $8 - 3(x - 4) = x - 20$

12) $5 + 2(x - 3) = 8$

Evaluate each expression using $a = \dfrac{3}{8}$ $b = \dfrac{2}{5}$ $n = \dfrac{3}{2}$ $x = 1\dfrac{2}{3}$ $w = \dfrac{5}{6}$

13) $bn + an$

14) $b[x - wn]$

15) $ab + x + w$

16) $n + b[n - a]$

17) Find the result when the square of $\frac{2}{3}$ is increased by the product of $1\frac{1}{6}$ and $\frac{4}{21}$.

18) Fannie Farkel's boat can go 7 MPH in still water. If she travels 2 hours and 15 minutes down a river whose current is $1\frac{2}{3}$ MPH, how far can she go?

19) Claudia brought a 20 pound watermelon to the Math Picnic. Boris ate $\frac{2}{5}$ of it, Olaf ate $\frac{1}{4}$ of it and Ferd Burfil ate $\frac{1}{10}$ of it. If Tarzan ate $\frac{3}{5}$ of the remainder, how much did he eat?

20) Egbert weighs $1\frac{1}{4}$ times the weight of his 120 pound sister, Wanda. Their Uncle Humperdink weighs $\frac{3}{5}$ of their combined weight. Find the weight of Uncle Humperdink.

CHAPTER 5
Equations with Fractions

In this chapter, you will learn to:
1. *Solve equations with fractions.*
2. *Solve proportion problems.*
3. *Solve word problems using proportions or equations with fractions.*

5.1 Solving equations with Fractions

In the last two chapters we have learned how to solve equations using integers and we have learned how to operate on fractions. We will now extend each of these skills as we solve equations using fractions.

The Reciprocal

In solving linear equations, we attempt to "undo" the equation. We would like to have 1x or 1y or 1z (or 1 of whichever variable is in the equation) and every thing else on the other side. Since we were limited to integers in Chapter 2, our equations took on a very limited form.

If $5x = 20$ and we want 1x, we divide each side by 5 to get $1x = 4$.

If $\frac{n}{4} = 24$ and we want 1n, we multiply each side by 4 to get $1n = 6$.

If $\frac{5}{6}y = 35$, how would we produce 1y? We multiply each side by 6 and divide each side by 5 to get 1y. This is the same as multiplying each side by the fraction $\frac{6}{5}$.

We call the fraction $\frac{6}{5}$ the reciprocal of $\frac{5}{6}$.

We will define **reciprocals** as two numbers whose product is 1. The number "1" is important in solving equations as our goal is to produce "1" of whatever variable is in the equation.

The reciprocal of $\frac{5}{8}$ is $\frac{8}{5}$ because $\frac{5}{8} \cdot \frac{8}{5} = 1$

The reciprocal of $\frac{-2}{3}$ is $\frac{-3}{2}$ because $\frac{-2}{3} \cdot \frac{-3}{2} = 1$

The reciprocal of 5 is $\frac{1}{5}$ because $5 \cdot \frac{1}{5} = 1$

Use of the Reciprocal in Solving Equations

The procedure used to solve equations having fractions will be the same procedure developed in Chapter 2 but will include the use of the reciprocal. This procedure is listed below:

1. Use distributive property and combining of like terms whenever possible.
2. Isolate the variable on one side of the equation (left or right.).
3. Produce a variable with a coefficient of 1. If the coefficient of the variable is a fraction, then we multiply each side of the equation by the reciprocal of the coefficient of the variable.

◆**Example 1**   Solve the equation: $\frac{4}{5}x + 2 = 26$

$\frac{4}{5}x + 2 - 2 = 26 - 2$ Subtract 2 to isolate the variable on left side

$\frac{4}{5}x = 24$ Combine like terms

$\frac{4}{5}x \cdot \frac{5}{4} = 24 \cdot \frac{5}{4}$ Multiply each side by the reciprocal of the coefficient of the variable

$1x = 30$ Produce 1x

◆**Example 2**   Solve the equation: $\frac{7}{8}x - 2 = \frac{1}{4}x + 4$

$\frac{7}{8}x - 2 - \frac{1}{4}x = \frac{1}{4}x + 4 - \frac{1}{4}x$ Isolate variable, subtract $\frac{1}{4}$ from each side

$\frac{7}{8}x - 2 - \frac{2}{8}x = 4$ Combine like terms, change to LCD

$\frac{5}{8}x - 2 + 2 = 4 + 2$ Add 2 to each side

$\frac{5}{8}x = 6$ Combine like terms

$\frac{5}{8}x \cdot \frac{8}{5} = 6 \cdot \frac{8}{5}$ Multiply each side by reciprocal of coefficient of variable

$1x = \frac{48}{5} \text{ or } 9\frac{3}{5}$ Produce 1x

Solving Word Problems

When an algebraic word problem contains fractions, the organization will be the same as we developed in Chapter 3 for word problems containing integers. The 5 step process is listed below:

1. Set up **categories** of information.
2. Assign **numbers** to the categories.
3. Write an **equation**.
4. **Solve** the equation.
5. Write the answer in the form of a **sentence**.

◆**Example 3** Use the 5 step process to solve the following problem:

Six more than $\frac{2}{3}$ of a number is 9. What is the number?

Let n = the number   **Category** of information and **number** with the category

$6 + \frac{2}{3}n = 9$ Translate to form an **equation**

$6 + \frac{2}{3}n - 6 = 9 - 6$ **Solve** the equation, subtract 6 from each side

$\frac{2}{3}n = 3$ Combine like terms

$\frac{2}{3}n \cdot \frac{3}{2} = 3 \cdot \frac{3}{2}$ Multiply each side by the coefficient of the variable

$1n = \frac{9}{2}$ or $4\frac{1}{2}$ Produce 1n

$\boxed{\text{The number is } 4\frac{1}{2}.}$ Write a **sentence**

◆**Example 4** Use the 5 step process to solve the following problem:

Find two numbers whose sum is eight such that three times the larger decreased by twice the smaller is five.

Let a = the larger number

8 − a = the smaller number   **Categories** and **numbers**

3a − 2(8 − a) = 5   Translate to form an **equation**

3a − 16 + 2a = 5   **Solve**

5a − 16 = 5

5a − 16 + 16 = 5 + 16

5a = 21

$5a \cdot \frac{1}{5} = 21 \cdot \frac{1}{5}$

$1a = \frac{21}{5}$ or $4\frac{1}{5}$

$8 - 4\frac{1}{5} = 3\frac{4}{5}$

$\boxed{\text{The larger number is } 4\frac{1}{5} \text{ and the smaller number is } 3\frac{4}{5}.}$ **Sentence**

Name:_____ 5.1 Problem Set

Solve each of the following equations.

1) $5 + 2(x - 3) = 8$

2) $6 - 3(x + 5) = 7$

3) $4 - 5(2x + 3) + 6x = 12$

4) $\frac{2}{3}x + 12 = 20$

5) $\frac{3}{8}x - 2 = 16$

6) $\frac{5}{6}x - 3 = 9$

7) $\frac{7}{8}x - 2 = \frac{3}{8}x + 5$

8) $\frac{5}{6}x - 3 = \frac{1}{3}x + 2$

9) $\frac{7}{12}x - 5 = \frac{1}{3}x + 2$

Evaluate each expression using $a = 2\frac{1}{3}$ $b = 1\frac{1}{2}$ $n = \frac{3}{4}$ $w = \frac{2}{3}$ $x = \frac{5}{6}$

10) $an + bx$

11) $b + n[\,b - x\,]$

12) $a + b[\,w - x\,]$

13) $n - a[\,n - b\,]$

Solve each of the following word problems.

14) When seven is added to four times a number the result is 25. Find the number.

15) Eight more than five times the sum of a number and two will be 32. Find the number.

16) Find two numbers whose sum is 10 and such that five times the smaller increased by twice the larger is 33.

17) Find two numbers whose sum is 8 such that six times the larger decreased by four times the smaller was 9.

18) For the Math BBQ party, Rod Stewart bought 20 pounds of meat. He bought $2\frac{1}{2}$ times as much beef as pork, and four more pounds of chicken than beef. How much of each type of meat did he buy?

19) Batman's Batboat can travel a steady seven miles per hour in still water. On one of his journeys to battle evil, he went up a river with a current of $1\frac{2}{3}$ MPH. How far did Batman travel if it took him exactly an hour and 12 minutes to catch the villain?

5.2 Using the Distributive Property on Equations with Fractions

In this section, we will continue to build our skill at working with fractions and solving equations with fractions. As an added feature, we will encounter equations that will be solved using the distributive property, described in Chapter 2. For review, we will state it here. If each of a, b, and c are real numbers the following is true:

$$a(b + c) = a \cdot b + a \cdot c \qquad \text{and} \qquad a(b - c) = a \cdot b - a \cdot c$$

We will now apply this property to equations with fractions.

◆**Example 1** Solve the equation.

$$3 + \frac{2}{3}(y+6) = 10$$

$$3 + \frac{2}{3}y + \frac{2}{3} \cdot 6 = 10$$

$$3 + \frac{2}{3}y + 4 = 10$$

$$\frac{2}{3}y + 7 = 10$$

$$\frac{2}{3}y + 7 - 7 = 10 - 7$$

$$\frac{2}{3}y = 3$$

$$\frac{2}{3}y \cdot \frac{3}{2} = 3 \cdot \frac{3}{2}$$

$$y = \frac{9}{2}$$

◆**Example 2** Solve the following equation.

$$5 + \frac{3}{4}(x-8) = 7$$

$$5 + \frac{3}{4}x - \frac{3}{4} \cdot 8 = 7$$

$$5 + \frac{3}{4}x - 6 = 7$$

$$\frac{3}{4}x - 1 = 7$$

$$\frac{3}{4}x - 1 + 1 = 7 + 1$$

$$\frac{3}{4}x = 8$$

$$\frac{3}{4}x \cdot \frac{4}{3} = 8 \cdot \frac{4}{3}$$

$$x = \frac{32}{3} \text{ or } 10\frac{2}{3}$$

◆**Example 3** Use the 5 step process to solve the following problem.

During the first hour of a cherry picking contest, Tarzan picked 20 pounds more than the amount picked by Batman. When the weight of Batman's cherries was added to $\frac{4}{5}$ the weight of Tarzan's cherries, the weight was exactly 28 pounds. Find the weight of the cherries picked by Tarzan.

w = weight of cherries picked by Batman

w + 20 = weight of cherries picked by Tarzan   **Categories** and **numbers**

$w + \frac{4}{5}(x + 20) = 28$ Translation into an **equation**

$w + \frac{4}{5} \cdot w + \frac{4}{5} \cdot 20 = 28$ **Solve** the equation

$\frac{9}{5}w + 16 = 28$

$\frac{9}{5}w + 16 - 16 = 28 - 16$

$\frac{9}{5}w = 12$

$\frac{9}{5}w \cdot \frac{5}{9} = 12 \cdot \frac{5}{9}$

$w = \frac{20}{3} \text{ or } 6\frac{2}{3}$

If $w = 6\frac{2}{3}$, then $w + 20 = 26\frac{2}{3}$

> Tarzan picked $26\frac{2}{3}$ pounds of cherries.   Write a **sentence**

Name:_____

5.2 Problem Set

Solve the following equations.

1) $\frac{2}{3}x + 5 = 7$	2) $\frac{5}{8}x + \frac{1}{4}x + 2 = 6$	3) $\frac{4}{5}x + 6 - \frac{1}{2}x = 14$
4) $8 + \frac{2}{3}(x - 9) = 18$	5) $5 + \frac{3}{4}(x + 6) = 7\frac{1}{2}$	6) $2\frac{1}{4} + \frac{2}{3}(x + 6) = 4\frac{1}{4}$
7) $3\frac{1}{3} + 2\frac{1}{2}(x - 6) = 1\frac{2}{3}$	8) $4 - \frac{2}{3}(x - 12) = 4$	9) $\frac{3}{4}(x + 12) + \frac{2}{3}(x - 9) = 37$

Solve the following word problems.

10) Miriam did some concrete work in her backyard. She used $2\frac{3}{4}$ cubic yards on a walk, $1\frac{1}{2}$ cubic yards on a slab for the hot tub, and $\frac{2}{3}$ cubic yards on the mowing strip. Her concrete company only sold in whole cubic yard amounts. If concrete sells for $ 37 per cubic yard, what did she pay for the concrete?

11) Baby Lisa weighed $6\frac{2}{3}$ pounds at birth. and 20 pounds on her first birthday. What was her average monthly gain during her first year of life?

12) Find two numbers whose sum is 8, such that five times the smaller diminished by three times the larger is six.

13) Find two numbers whose sum is 9, such that four times the larger increased by twice the smaller is 29.

14) One number is three more than twice another. Five times the smaller added to three times the larger is 15. Find the larger number.

15) For the Math party, Wolfgang bought one pound more than twice as much chicken as beef. If he bought a total of 15 pounds of meat, how much chicken did he buy?

16) Find two numbers whose sum is three if five times the larger decreased by twice the smaller is 9.

17) Mrs. Garcia weighs exactly $2\frac{1}{2}$ times the weight of her daughter Lucy. Together they weigh 161 pounds. Find the weight of Batwoman who is seven pounds heavier than Mrs. Garcia.

5.3 Ratios and Proportions

In this section we will discuss fractions using the word ratio. Ratios are often used to compare quantities, and as we shall see, there are many situations in which the concept of ratio is used to describe everyday situations. Miles per gallon, wage earned per hour and cost per ounce of cereal are a few examples.

What is a Ratio ?

A **ratio** is the comparison of two quantities and is usually written as a fraction. Symbolically,

if a and b are real numbers, then the ratio of a to b is $\frac{a}{b}$, $b \neq 0$

The order of comparing quantities does make a difference. For example, if a math class had 20 women and 10 men, then we could express the ratio of women to men as:

20 women to 10 men and write: $\frac{20}{10}$ or $\frac{2}{1}$ as the reduced form.

We could express the ratio of men to women as:

10 men to 20 women and write: $\frac{10}{20}$ or $\frac{1}{2}$ as the reduced form.

We could express the ratio of men to total number of students as:

10 men to 30 students and write: $\frac{10}{30}$ or $\frac{1}{3}$ as the reduced form.

In each of the above examples we get radically different answers by comparing the quantities in different order. Notice that the first number given in the comparison is the numerator of the ratio and the second number given is the denominator of the ratio. As with fractions in general, the reduced form is usually the preferred form.

◆**Example 1** Express the ratio of 25 to 35 as a fraction reduced to lowest terms.

$$\frac{25}{35} = \frac{5}{7}$$

Notice that the **pattern** in the translation places the first quantity 25 in the numerator and the second quantity 35 in the denominator.

◆**Example 2** Express the ratio of 36 to 12 as a fraction reduced to lowest terms.

$$\frac{36}{12} = \frac{3}{1}$$

Although we know that $\frac{3}{1} = 3$, we still keep the 1 in the denominator and state that the ratio is 3 to 1. This is sometimes written as 3:1

◆**Example 3** Express the ratio of $\frac{2}{3}$ to $\frac{5}{9}$ as a fraction reduced to lowest terms.

$$\frac{\frac{2}{3}}{\frac{5}{9}} = \frac{2}{3} \div \frac{5}{9}$$

$$= \frac{2}{3} \cdot \frac{9}{5}$$

$$\boxed{= \frac{6}{5}}$$

◆**Example 4** Express the ratio of $1\frac{1}{5}$ to $2\frac{2}{3}$ as a fraction reduced to lowest terms.

$$\frac{1\frac{1}{5}}{2\frac{2}{3}}$$

$$= 1\frac{1}{5} \div 2\frac{2}{3}$$

$$= \frac{6}{5} \div \frac{8}{3}$$

$$= \frac{6}{5} \cdot \frac{3}{8}$$

$$\boxed{= \frac{9}{20}}$$

Ratio as a Rate

When a ratio is used to compare different quantities the ratio is called a **rate**. For example, the gas efficiency of a car is often given in miles per gallon (MPG). If a car traveled 210 miles on 7 gallons of gas, we could say,

$$\frac{210 \text{ miles}}{7 \text{ gallons}} = \frac{30 \text{ miles}}{1 \text{ gallon}}$$

The car burns gas at the rate of 30 miles per gallon (30 MPG).

◆**Example 5** If a car traveled 160 miles in 4 hours, what is the car's rate in miles per hour?

The ratio of miles to hours is $\frac{160 \, miles}{4 \, hours} = \frac{40 \, miles}{1 \, hour}$

The average rate then is 40 miles per hour.

◆**Example 6**   A 120 gallon tank was filled in 15 minutes. What was the rate in gallons per minute?

$$\text{The ratio of gallons to minutes is } \frac{120 \text{ gallons}}{15 \text{ minutes}} = \frac{8 \text{ gallons}}{1 \text{ minute}}$$

$$= 8 \frac{\text{gallons}}{\text{minute}}$$

> The average rate then is 8 gallons per minute.

Unit Pricing as a Rate

Smart shoppers often use a type of rate called unit pricing to compare the same item packaged in different size containers. **Unit pricing** is the ratio of the price to the quantity. When reduced, the ratio gives price per one unit.

◆**Example 7**   The book store sells the same type of paper in two different packages. One package has 40 sheets and sells for $2 and the other package has 200 sheets and sells for $6. Use unit pricing to determine the better buy.

Using the unit pricing ratio, $\frac{Price}{Quantity}$ we can compare the two packages.

First package: $\dfrac{\$2}{40 \, sheets} = \dfrac{200¢}{40 \, sheets}$

$= \dfrac{5¢}{1 \, sheet}$ The unit price is 5¢ per sheet.

Second package: $\dfrac{\$6}{200 \, sheets} = \dfrac{600¢}{200 \, sheets}$

$= \dfrac{3¢}{1 \, sheet}$ The unit price is 3¢ per sheet.

> The second package would be the better buy.

Proportions

A **proportion** is the equality of two ratios. For example the ratio of 3 to 4 is the same as the ratio of 15 to 20, so we say they are proportional. Symbolically we say,

if the ratio $\frac{a}{b}$ has the same value as $\frac{c}{d}$ then the equality, $\frac{a}{b} = \frac{c}{d}$ is called a proportion. (b ≠ 0 and d ≠ 0).

In the proportion, $\frac{a}{b} = \frac{c}{d}$

    a is called the **first** term of the proportion

    b is called the **second** term of the proportion

    c is called the **third** term of the proportion

    d is called the **fourth** term of the proportion

Let us now look for a **pattern** in proportions:

$\frac{3}{4} = \frac{6}{8}$ The product of the first and fourth terms is 3 • 8 = **24**

           The product of the second and third is 4 • 6 = **24**

$\frac{5}{8} = \frac{10}{16}$ The product of the first and fourth terms is 5 • 16 = **80**

           The product of the second and third is 8 • 10 = **80**

$\frac{4}{5} = \frac{12}{15}$ The product of the first and fourth terms is 4 • 15 = **60**

           The product of the second and third is 5 • 12 = **60**

The pattern appears that in each proportion the product of the first and fourth is the same as the product of the second and third. This pattern is so useful that we give special names to these terms. The first and fourth terms are called the **means** and the second and third terms are called the **extremes**. Mathematicians have a theorem for this pattern which is given below:

For any proportion, the product of the means equals the product of the extremes. Symbolically we write:

If $\frac{a}{b} = \frac{c}{d}$ then b • c = a • d

This theorem will have immediate use as we solve proportion problems.

Section 5.3: Ratios and Proportions

◆ **Example 8** Find the value of x if $\dfrac{x}{6} = \dfrac{15}{18}$

$\qquad 6 \cdot 15 = 18x \qquad$ The product of the means = product of the extremes

$\qquad 90 = 18x \qquad$ Solve the equation

$\qquad \dfrac{90}{18} = \dfrac{18x}{18}$

$\qquad \boxed{x = 5}$

◆ **Example 9** Find the value of n if $\dfrac{8}{n} = \dfrac{6}{5}$

$\qquad 6n = 40 \qquad$ The product of the means = the product of the extremes

$\qquad \dfrac{6n}{6} = \dfrac{40}{6} \qquad$ Solve the equation

$\qquad \boxed{n = \dfrac{20}{3}}$

◆ **Example 10** Find the value of w if $\dfrac{1\frac{4}{5}}{10} = \dfrac{\frac{3}{4}}{w}$

$\qquad 1\dfrac{4}{5} \cdot w = \dfrac{3}{4} \cdot 10$

$\qquad \dfrac{9}{5} \cdot w = \dfrac{15}{2}$

$\qquad \dfrac{9}{5} \cdot w \cdot \dfrac{5}{9} = \dfrac{15}{2} \cdot \dfrac{5}{9}$

$\qquad \boxed{w = \dfrac{25}{6}}$

Name:_____   5.3 Problem Set

Express each of the following as a ratio reduced to its lowest terms.

1) 12 to 18	2) 28 to 35	3) 15 to 6
4) $\frac{2}{3}$ to $1\frac{1}{4}$	5) $3\frac{1}{2}$ to $1\frac{5}{6}$	6) $2\frac{1}{3}$ to $4\frac{2}{3}$

Alfonso made a fruit salad using the following ingredients: 4 pounds of oranges, 2 pounds of bananas, 3 pounds of strawberries and 6 pounds of apples. Use this information to express the ratio of each of the following.

7) Oranges to bananas	8) Strawberries to apples	9) Oranges to apples
10) Strawberries to the salad	11) Apples to the salad	12) Oranges and apples to the salad

Express each of the following as a rate.

13) 105 miles in 3 hours	14) 68 gallons in 4 hours	15) 276 miles using 12 gallons

Find the unit price of each of the following.

16) $102 for 17 pounds	17) $252 for 14 kilograms	18) $138 for 46 yards

Solve each of the following proportion problems.

19) $\dfrac{2}{3} = \dfrac{18}{x}$

20) $\dfrac{a}{8} = \dfrac{9}{4}$

21) $\dfrac{5}{w} = \dfrac{8}{3}$

22) $\dfrac{7}{8} = \dfrac{n}{5}$

23) $\dfrac{4}{y} = \dfrac{5}{6}$

24) $\dfrac{x}{8} = \dfrac{5}{12}$

25) $\dfrac{\frac{3}{5}}{b} = \dfrac{\frac{3}{4}}{15}$

26) $\dfrac{\frac{2}{5}}{8} = \dfrac{x}{10}$

27) $\dfrac{4\frac{1}{2}}{12} = \dfrac{2\frac{1}{3}}{a}$

28) $\dfrac{1\frac{3}{4}}{10} = \dfrac{2\frac{4}{5}}{y}$

29) $\dfrac{6}{1\frac{3}{5}} = \dfrac{a}{3\frac{1}{3}}$

30) $\dfrac{2\frac{2}{5}}{1\frac{1}{6}} = \dfrac{18}{z}$

Solve the following equations.

31) $\frac{5}{8}x + 7 = 22$

32) $\frac{5}{6}x - 3 = \frac{1}{3}x + 5$

33) $\frac{3}{5}(a+20) = \frac{1}{5}a + 15$

34) $\frac{8}{3}(n-6) = \frac{2}{3}(n+24)$

35) $\frac{1}{2}(w+8) = \frac{1}{3}(w+18)$

36) $\frac{1}{4}(x-24) = \frac{1}{5}(x+20)$

5.4 Word Problems Using Proportions

In this section, we will apply the concepts of proportion to the solution of word problems.

Word Problem Organization

The organization for solving a proportion type word problem will be the same as or any other algebraic word problem. A review of the 5 step process is listed below:

1. **Categories** of information.
2. **Numbers** with the categories.
3. Form an **equation**.
4. **Solve** the equation.
5. Answer the question with a **sentence**.

Helpful Hint 11: Be careful to form the proportion in the correct order. For example, if the left hand ratio compares miles to gallons, then the right hand ratio must compare miles to gallons.

♦ Example 1 On the first 300 miles of their vacation trip, the Chan family used 12 gallons of gas. If they continued using gas in the same ratio, how much gas would they use driving their entire 1000 mile trip?

Let x = amount of gas used driving 1000 miles

$$\frac{x}{1000} = \frac{12}{300} \qquad \frac{gas}{miles} = \frac{gas}{miles}$$

$$300x = 12,000$$

$$\frac{300x}{300} = \frac{12,000}{300}$$

$$x = 40$$

The Chan car would use 40 gallons of gas in 1000 miles.

◆**Example 2** Gilda read 60 pages in her history book in 2 hours. If she continued reading at the same rate, what would her total time be to read her full 100 page assignment in history?

Let h = number of hours needed to read the whole assignment

$$\frac{h}{100} = \frac{2}{60} \qquad \frac{hours}{pages} = \frac{hours}{pages}$$

$$60h = 200$$

$$\frac{60h}{60} = \frac{200}{60}$$

$$h = \frac{10}{3} \text{ or } 3\frac{1}{3}$$

Gilda would read the 100 page assignment in $3\frac{1}{3}$ hours.

◆**Example 3** How much sugar is in 28 oz. cereal if there is 3 oz. of sugar in 8 oz. of the same type cereal?

Let a = amount of sugar in the 28 oz. of cereal

$$\frac{a}{28} = \frac{3}{8} \qquad \frac{sugar}{cereal} = \frac{sugar}{cereal}$$

$$8a = 84$$

$$\frac{8a}{8} = \frac{84}{8}$$

$$a = \frac{21}{2} \text{ or } 10\frac{1}{2}$$

There is $10\frac{1}{2}$ oz. of sugar in 28 oz. of this type cereal.

Helpful Hint 12: *As a reminder that your proportion is set up correctly on each side of the equal sign, write the proportion in words next to the proportion written in numbers as in the three examples above.*

Name:_____   5.4 Problem Set

Express each of the following as a ratio.

1) 14 to 35	2) 125 to 25	3) 45 to 20
4) $\frac{3}{4}$ to $\frac{5}{8}$	5) $\frac{7}{8}$ to $3\frac{1}{4}$	6) $5\frac{1}{4}$ to $4\frac{3}{8}$

Solve each of the following proportions.

7) $\frac{8}{15} = \frac{x}{40}$	8) $\frac{10}{n} = \frac{4}{15}$	9) $\frac{w}{8} = \frac{35}{56}$
10) $\frac{7}{12} = \frac{21}{a}$	11) $\frac{9}{x} = \frac{12}{5}$	12) $\frac{n}{16} = \frac{5}{4}$
13) $\frac{5}{7} = \frac{x}{3\frac{1}{2}}$	14) $\frac{3\frac{1}{2}}{n} = \frac{14}{3}$	15) $\frac{3\frac{1}{2}}{a} = \frac{1\frac{2}{5}}{4\frac{1}{2}}$

Solve the following proportion problems.

16) The ratio of men to women in a certain school is 5 to 4. If there are 920 men in the school, how many women are there?

17) Wilma bought a $16,000 car to celebrate her 27th birthday. During the first 3 months, she drove a total of 4000 miles. If she continues to drive at this rate, how many miles would she have driven at the end of two years?

18) The Ferd Burfil Golf Ball Company found that during an average run, there were 9 defective balls out of each 800 balls produced. How many defective balls could be expected out of a production of 25,600 balls?

19) A mathematics teacher can grade 5 tests in 21 minutes. If she continues at this rate, how many hours will she spend to grade a set of 35 tests?

Solve the following problems.

20) Find the unit price for a computer chip if 240 chips cost $4080.

21) What is the unit price for a cubic yard of concrete if 28 cubic yards sell for $1148?

22) If a 345 gallon container can be filled in 15 minutes, what is the rate at which the container is filled?

23) A factory produces 224 computers during each working shift of 8 hours. At what rate does this factory build computers?

24) $\frac{3}{5}(w-35) = \frac{1}{5}w + 5$

25) $\frac{7}{3}(x-6) = \frac{5}{3}(x+9)$

5.5 Chapter Review

In this chapter you have learned to :

1. Solve equations with fractions.
2. Solve proportion problems.
3. Solve word problems using proportions or equations with fractions.

How to Solve an Equation with Fractions   [5.1]

To solve equations with fractions:

1. Use distributive property and combine like terms whenever possible.
2. Isolate the variable on one side of the equation (left or right).
3. Produce a variable with a coefficient of 1. If the coefficient of the variable is a fraction, multiply each side of the equation by the **reciprocal** of the coefficient of the variable.

◆ **Example 1**   Solve the following equation: $\frac{3}{4}(x+12) = \frac{2}{3}x + 13$

$$\frac{3}{4}x + \frac{3}{4} \cdot 12 = \frac{2}{3}x + 13 \qquad \textit{Distribute}$$

$$\frac{3}{4}x + 9 = \frac{2}{3}x + 13$$

$$\frac{3}{4}x + 9 - \frac{2}{3}x = \frac{2}{3}x + 13 - \frac{2}{3}x \qquad \textit{Isolate variable}$$

$$\frac{3}{4}x\left(\frac{3}{3}\right) + 9 - \frac{2}{3}x\left(\frac{4}{4}\right) = 13$$

$$\frac{9}{12}x + 9 - \frac{8}{12}x = 13$$

$$\frac{1}{12}x + 9 - 9 = 13 - 9$$

$$\frac{1}{12}x = 4$$

$$\frac{1}{12}x \cdot (12) = 4 \cdot (12) \qquad \textit{produce } 1x$$

$$\boxed{x = 48}$$

Ratios [5.3]

A ratio is the comparison of two quantities. A ratio is usually expressed as a common fraction reduced to lowest terms.

◆**Example 2** Express the ratio of 24 to 36 as a fraction reduced to lowest terms.

$$\frac{24}{36} = \frac{2}{3}$$

◆**Example 3** Express the ratio of $3\frac{1}{2}$ to $1\frac{1}{4}$ as a fraction reduced to lowest terms.

$$\frac{3\frac{1}{2}}{1\frac{1}{4}} = 3\frac{1}{2} \div 1\frac{1}{4}$$

$$= \frac{7}{2} \div \frac{5}{4}$$

$$= \frac{7}{2} \cdot \frac{4}{5}$$

$$\boxed{= \frac{14}{5}}$$

◆**Example 4** If a car traveled 224 miles in 4 hours what is the car's rate in miles per hour (MPH).

$$\frac{224 \ miles}{4 \ hours} = \frac{56 \ miles}{1 \ hour}$$

$$\boxed{\text{The car traveled an average rate of 56 MPH.}}$$

Word problems Using Equations with Fractions [5.2]

To solve an algebraic word problem using fractions use the procedure reviewed below:

1. **Categories** of information.
2. **Numbers** with the categories.
3. Form an **equation**.
4. **Solve** the equation.
5. Write the answer in a **sentence**.

◆**Example 5** One number is four more than $\frac{2}{3}$ of another number. If their sum is six what are the numbers?

Let n = first number

$$\frac{2}{3}n + 4 = the\ sec ond\ number$$

$$n + \frac{2}{3}n + 4 = 6$$

$$\frac{5}{3}x + 4 = 6$$

$$\frac{5}{3}x + 4 - 4 = 6 - 4$$

$$\frac{5}{3}n = 2$$

$$\frac{5}{3}n \cdot \left(\frac{3}{5}\right) = 2 \cdot \left(\frac{3}{5}\right)$$

$$n = \frac{6}{5}\ or\ 1\frac{1}{5} \qquad The\ first\ number$$

$$\frac{2}{3}n + 4 = \frac{2}{3}\left(\frac{6}{5}\right) + 4 \qquad The\ sec ond\ number$$

$$= \frac{4}{5} + 4$$

$$\boxed{The\ numbers\ are\ 1\frac{1}{5}\ and\ 4\frac{4}{5}}$$

Proportions   [5.3]

A proportion is the equality of two ratios. To solve a proportion use the theorem, "The product of the means = the product of the extremes." This is referred to as the "cross product" theorem, since lines drawn through the means and the extremes form a cross.

◆**Example 6**   Solve the proportion. $\dfrac{w}{6} = \dfrac{2}{5}$

$5w = 12$ The product of the means = the product of the extremes

$\dfrac{5w}{5} = \dfrac{12}{5}$

$$\boxed{w = \dfrac{12}{5} \text{ or } 2\dfrac{2}{5}}$$

Word Problems Using Proportions   [5.4]

The organization described above is also used for proportion type word problems. The formation of the equation (step 3) is the formation of the proportion. Remember to form the proportion in the correct order. For example if the left side of the proportion compares number of pounds to cost, the right side should also compare number of pounds to cost.

◆**Example 7**   Jung Han bought 5 pounds of strawberries for $8. Jung went home to taste the strawberries and found them so tasty that he went back to the store and bought 60 pounds at the same price and brought them to the Math Picnic. How much did Jung pay for the 60 pounds of strawberries?

Let a = the amount spent on 60 pounds of strawberries

$$\dfrac{\text{number of pounds}}{\text{total cost}} = \dfrac{\text{number of pounds}}{\text{total cost}}$$

$$\dfrac{5}{8} = \dfrac{60}{a}$$

$$5a = 8 \cdot 60$$

$$\dfrac{5a}{5} = \dfrac{480}{5}$$

$$a = 96$$

$\boxed{\text{Jung paid \$96 for the 60 pounds of strawberries.}}$

Name:_____

5.5 Chapter Review Test

Express each of the following as ratios reduced to lowest terms

1) 32 to 48	2) 175 to 25	3) $3\frac{1}{2}$ to $1\frac{1}{4}$

A serving of a popular cracker contains 2 grams of protein, 10 grams of carbohydrate and 6 grams of fat. Use this information to find the following ratios.

4) Protein to fat	5) Fat to carbohydrate	6) Carbohydrate to protein

A coin collection has 8 dimes, 12 nickels, 20 quarters and 16 pennies. Use this information to find the following ratios.

7) Dimes to nickels	8) pennies to dimes	9) Quarters to nickels
10) Nickels to quarters	11) Dimes to all of the coins	12) Quarters to pennies and dimes

Express each of the following as a rate.

13) 336 miles using 16 gallons	14) 84 miles in 6 hours	15) Type 325 words in 5 minutes

Express each of the following using unit pricing.

16) 23 pounds for $299	17) $720 for 16 boxes	18) 3 ounces for $1068

Solve the following proportions.

19) $\dfrac{x}{12} = \dfrac{42}{72}$	20) $\dfrac{9}{a} = \dfrac{6}{7}$	21) $\dfrac{4}{15} = \dfrac{w}{10}$
22) $\dfrac{3\frac{1}{2}}{5} = \dfrac{3}{x}$	23) $\dfrac{n}{2\frac{1}{2}} = \dfrac{6}{7}$	24) $\dfrac{x}{1\frac{1}{4}} = \dfrac{2\frac{1}{2}}{2\frac{1}{4}}$

Solve the following equations.

25) $\dfrac{5}{16}y + 8 = 38$

26) $\dfrac{3}{4}x - 2 = \dfrac{1}{8}x + 4$

27) $\dfrac{5}{6}(a-12) = 3$

28) $\dfrac{5}{6}(x-24) - \dfrac{1}{4}(x+16) = 4$

29) $\dfrac{7}{8}(x+16) = \dfrac{1}{4}(x+32)$

30) $\dfrac{3}{8} + \dfrac{5}{4}\left(x+\dfrac{1}{2}\right) = 3$

Solve the following word problems in the usual manner.

31) A car dealer orders 7 white cars for each 3 blue one. One year she ordered 651 white cars. How many blue cars did she order?

32) If the ratio of men to women at the college is 4 to 5 and the college has 11,460 men, how many women are in the college?

33) If a fruit stand sells 8 crates of avocados every 5 days, how many days will it take to sell 72 crates of avocados?

34) When 2 is added to $\frac{3}{4}$ of the sum of a number and 8 the result is 10. Find the number.

CHAPTER 6
Percent

In this chapter, you will learn to:

1. *Approximate answers for problems involving the four basic arithmetic operations, powers or square roots.*
2. *Use a hand held-calculator to perform the four basic arithmetic operations, powers or square roots.*
3. *Solve percentage problems.*
4. *Solve word problems involving percentage.*

6.1 Decimals and the Use of a Calculator

And now for something really useful - your calculator. This text does not include instruction on decimals because of the availability of the hand-held calculator and the ease of doing decimal calculations using it. From this point on, you will be expected to use your calculator whenever computations deserve its use.

Standard Keys

There are many different calculators available on the market today but most of them have the standard keys listed below. At this time be sure you can locate each of the following keys on your calculator. If you can not, please refer to your calculator instruction manual or ask someone for help.

Key	Description	Key	Description
$\boxed{x}$	Multiplication key	$\boxed{\sqrt{}}$	Square root key
$\boxed{+}$	Addition key	$\boxed{x^2}$	Square key
$\boxed{\div}$	Division key	$\boxed{+/-}$	Sign change key
$\boxed{-}$	Subtraction key	$\boxed{\frac{1}{x}}$	Reciprocal key
$\boxed{=}$	Equals key	$\boxed{(}$	Left parenthesis
$\boxed{.}$	Decimal point	$\boxed{)}$	Right parenthesis

$\left.\begin{array}{l}\boxed{\text{2nd}} \\ \boxed{\text{INV}} \\ \boxed{\text{SHIFT}}\end{array}\right\}$ These keys activate the second function on each key.

$\left.\begin{array}{l}\boxed{x^y} \\ \boxed{y^x}\end{array}\right\}$ General exponent keys

The explanations and examples which follow assume that you are using a scientific calculator with algebraic logic. For calculators with algebraic logic, the rules for order of operations are built into the machines. (If your calculator features Reverse Polish Notation, you will have to refer to the instruction that came with your calculator.) As there are many different calculators available today, yours might be different than the one described here. You might have to experiment with yours to get the correct operating procedure. If you are still having trouble, ask your teacher or some classmate for help. Always keep your instruction manual handy, as each different calculator model has unique features which might not be familiar to your teacher or your friend.

◆**Example 1** Use your calculator to find: $5.2 + (13)(7.6) = $ **104**

Sequence of keys: [5] [.] [2] [+] [13] [x] [7.6] [=] **display 104**

|NOTE| In the above example, the number 5.2 was done with step-by-step key strokes, while the number 7.6 was done assuming that you know how to punch the key strokes for a decimal number. The same assumption will be made on all of the succeeding examples.

◆**Example 2** Use your calculator to find: $8 + 3[5.9 + 6.3] = $ **44.6**

Sequence of keys: [8] [+] [3] [x] [[] [5.9] [+] [6.3] []] [=] **display 44.6**

The Square Root Key

We have often discussed the squaring of a number as repeated multiplication. For example $5^2 = 25$, because $5 \cdot 5 = 25$. We will now define the square root function as the inverse function. The **square root** of a non-negative number x, written $\sqrt{x}$, is the number that we square to obtain x. Symbolically we write:

$$\sqrt{x^2} = x, \text{ if x is greater than or equal to zero}$$

Examples:
$\sqrt{49} = 7$ because $7^2 = 49$
$\sqrt{36} = 6$ because $6^2 = 36$
$\sqrt{81} = 9$ because $9^2 = 81$

The above examples can be done easily because the numbers are small. When the numbers get larger and not so familiar, the **square root key** is a fast way to get an answer. Use your square root key to verify that each of the following is correct: You will notice that the square function and the square root function occupy the same key, but to access the square function, you must first press the inverse or second function key.

$\sqrt{784} = 28$
$\sqrt{2809} = 53$
$\sqrt{5476} = 74$

◆**Example 3** Use your calculator to find: $48 + 36\sqrt{529} \div 18 = $ **94**

Sequence of keys: [48] [+] [36] [x] [529] [√] [÷] [18] [=] **display 94**

◆**Example 4** Use your calculator to find: $24^2 - 5\sqrt{961} = 421$

Sequence of keys: $\boxed{24}$ $\boxed{\text{inv}}$ $\boxed{\sqrt{}}$ $\boxed{-}$ $\boxed{5}$ $\boxed{\times}$ $\boxed{961}$ $\boxed{\sqrt{}}$ $\boxed{=}$ **display 421**

◆**Example 5** Use your calculator to find: $\dfrac{24 + 46}{(2)(7)} = 5$

Sequence of keys: $\boxed{(}$ $\boxed{24}$ $\boxed{+}$ $\boxed{46}$ $\boxed{)}$ $\boxed{\div}$ $\boxed{(}$ $\boxed{2}$ $\boxed{\times}$ $\boxed{7}$ $\boxed{)}$ $\boxed{=}$ **display 5**

The Reciprocal Key

In Section 5.1, we defined reciprocals as two numbers whose product is 1.

The reciprocal of 3 is $\dfrac{1}{3}$. The reciprocal of 8 is $\dfrac{1}{8}$. The reciprocal key for any number x is $\dfrac{1}{x}$.

◆**Example 6** Use your calculator to find: $\dfrac{1}{5} + 18^2 = 324.2$

Sequence of keys: $\boxed{5}$ $\boxed{\tfrac{1}{x}}$ $\boxed{+}$ $\boxed{18}$ $\boxed{\text{INV}}$ $\boxed{\sqrt{}}$ $\boxed{=}$ **display 324.2**

Approximating Answers

Whenever you do any kind of computation, on a calculator or otherwise, it is always a good idea to have in mind an **approximate** answer for whatever the problem. You probably do this mental exercise whenever you go shopping. For example, if you go to the store and buy 12 cans of soda and three bags of chips for a math picnic, you would probably approximate that the cost would be about $10. Depending on where you shop the price could be more or it could be less. If, however, the clerk hit the wrong key on the cash register, and wanted you to pay $64.92, you would likely know there is some kind of error. You are expected to seek the same kind of reasonable answers as you work problems in this book. For each problem that you do, ask yourself if the answer you are getting is a reasonable one.

Helpful Hint 13: To approximate answers to arithmetic calculations, use easy to calculate numbers rather than the given numbers of the problem. For example, to multiply (4.8) • (8.9), use the values of (5) • (9) = 45 as the approximate values.

◆**Example 7** Approximate and then use a calculator to find: $98.3 + 5\sqrt{52.67}$

$$\text{Approximation: } 98.3 + 5\sqrt{52.67} \approx 100 + 5(7)$$

$$\approx 100 + 35$$

$$\approx 135 \quad \text{Approximate answer}$$

$$\text{Calculator} \quad 98.3 + 5\sqrt{52.67} = 134.58705$$

The approximated answer is reasonably close to the calculator answer.

Chapter 6: Percent

◆ Example 8 Approximate and then use a calculator to find: $(4.8)^2 - (3.7)(2.1)$

Approximation: $(4.8)^2 - (3.7)(2.1)$

$$\approx 5^2 - (4)(2)$$
$$\approx 25 - 8$$
$$\approx 17 \quad \text{Approximate answer}$$

Calculator: $(4.8)^2 - (3.7)(2.1) = \mathbf{15.27}$

Again, the answer seems reasonable.

◆ Example 9 Approximate and then use calculator to find $8.1\sqrt{37.2} + (3.9)^3$

Approximation: $8.1\sqrt{37.2} + (3.9)^3$

$$\approx 8\sqrt{36} + (4)^3$$
$$\approx (8)(6) + 64$$
$$\approx 48 + 64$$
$$\approx 112 \quad \text{Approximate answer}$$

Calculator: $8.1\sqrt{37.2} + (3.9)^3 = \mathbf{108.72236}$

The answer seems reasonable.

Place Value

The place that a digit occupies relative to the decimal point determines the digit's **place value**. The whole number values are to the left of the decimal point; the fractional values are to the right of the decimal point. Since we use base 10 as our number system, the fractional parts as well as the whole number parts have a place value as some power of 10. Some fractional equivalents are given below:

$$\frac{1}{10} = 0.1 \text{ and is read as "one - tenth"}$$

$$\frac{1}{100} = 0.01 \text{ and is read as "one - one hundredths"}$$

$$\frac{1}{1000} = 0.001 \text{ and is read as "one - one thousandths"}$$

The following table summarizes some of the place value locations:

hundreds	tens	ones	Dec. Point	tenths	hundredths	thousandths	ten thousandths
100	10	1	.	0.1	0.01	0.001	0.0001

◆ Example 10 Use place value to write the number 486.032

Four hundred eighty-six and thirty-two thousandths.

Rounding Numbers

There are times when its a good idea to round off numbers. In measuring distances between major cities, it is enough to know a rounded off number like the distance from San Francisco to Los Angeles might be given as 400 miles, rounded to the nearest hundred miles If, on the other hand, you needed directions to someone's house within a city, you might like distances given accurate to the nearest tenth of a mile. For example, the house is "2.4 miles past the second signal light after the exit from the freeway." Population figures are often given in rounded off numbers. The population of the world is five billion people rounded off to the nearest billion.. While the rounded off numbers may seem somewhat non-precise, the way that one rounds off numbers does follow a precise method.

General Round off Procedure

1. Locate the digit occupying the **place value** to which you will round the number.

2. Locate the next digit to the right, which we will call the **critical digit.**

3. a) If the **critical digit** is five or greater (5,6,7,8 or 9), round up by adding 1 to the digit occupying the place value to which you are rounding and replace the **critical digit** and any other digits to its right with zeros. (Delete ending zeros to the right of the decimal point. For example, write 4.35 instead of 4.350.)

   b) If the **critical digit** is less than 5 (0,1,2,3, or 4), round down by replacing the **critical digit** and all digits to its right with zeros.

◆**Example 11**   Round 345.684 to nearest to the nearest **tenths.**

The digit **6** occupies the tenths place, so the digit, **8** is the **critical digit.** Since 8 is greater than 5, add 1 to the **6** and all the digits to the right are replaced by zeros, but since the zeros are to the right of the decimal point we will delete them.

> The number 345.684 rounded to nearest tenth is 345.7

◆**Example 12**   Round 23.614 to the nearest **hundredths.**

The digit **1** occupies the hundredths place, so the digit, **4 is the critical digit.** Since 4 is less than 5, replace the 4 with 0. but since the 0 is to the right of the decimal point delete it as well.

> The number 23.614 rounded to nearest hundredths is 23.61

◆**Example 13**   Round 582.36 to the nearest **hundreds**

> The number 582.36 rounded to nearest hundred is 600

Degree of Accuracy

When measurements are taken, there are times when great precision is needed and accuracy is given to the nearest hundredths of a centimeter or nearest millionths of a centimeter. In particular, in the fields of chemistry, biology and physics when very small objects are to be measured. At other times, whole numbers rounded to the nearest hundred or nearest million will be adequate. A golfer, for example would only be interested in yardage given in whole numbers; the federal budget figures are often given with numbers rounded to nearest million are even nearest billion. Most often the context of the measurement should give a clue as to the place value to which the numbers should be rounded. When you are looking only at a number, the number of place-value positions filled determines the decimal precision of a given measurement. For example a measurement of 12.7 feet indicates that the measurement has been made accurate to the nearest tenth of a foot. A measurement of 2.76 meters means that the measurement has been made accurate to the nearest hundredths of a meter.

Rule of Thumb

Most often all measurements for a single object are made with the same degree of accuracy. That is if we are going to measure a room, we decide ahead of time the precision needed for the specific purpose. If we are measuring the floor for the installation of a carpet, we need more accuracy than if we are measuring the same floor to paint it. If someone supplies us with some measurements with differing degrees of accuracy, the **rule of thumb** is to round the final calculation to the **precision of the least precise** number involved in the computation.

◆ **Example 14** Find the area of a rectangle whose measurements are given to be 12.3 feet in length and 8.65 feet in width.

$$\text{Using the formula,} \quad A = lw$$

$$A = (12.3 \text{ ft})(8.65 \text{ ft})$$

$$A = 106.395 \text{ ft}^2$$

Which rounds off to be 106.4 ft^2 because the **least precise** measurement, 12.3 ft was accurate to the nearest tenth.

◆ **Example 15** Find the product of 74 and the square of 3.8.

This translates to be $74(3.8)^2 = 1068.56$ but will round off to be 1069 because the least accurate number, 74, is only given to whole number accuracy.

Computations involving money

Another rule of thumb deals with **computations involving money**. Since the smallest coin in our country is a penny, final answers will round to the nearest cent, which means to the **nearest hundredths**.

◆**Example 16** A carpenter agrees to build a deck and charges $12.60 per square foot. The deck measures 23.4 feet by 16.7 feet. Find the cost of the deck.

First find out the **area** using the formula $A = lw$

$$A = (23.4 \text{ ft})(16.7 \text{ ft})$$

$$A = 390.78 \text{ ft}^2$$

Now the **cost**,   cost $= (390.78 \text{ ft}^2)(\$12.60)$

cost $= \$4923.828$

Now round off to the nearest cent.

> The deck would cost $4923.83

Mathematical translation of the word "of"

The word "of" sometimes translates into the operation of multiplication.

The statement "$\frac{2}{3}$ of x" means $\frac{2}{3} \cdot x$ or just $\frac{2}{3} x$

The statement "$\frac{1}{2}$ of the cost" means $\frac{1}{2}$ times the cost.

◆**Example 17** Find $\frac{3}{4}$ of 84.

$\frac{3}{4} \cdot 84$ Translation using the word "of"

$\frac{3}{4} \cdot 84 = 63$

◆**Example 18** $\frac{5}{6}$ of what number is 35 ?

Let x = the number

$\frac{5}{6} x = 35$ Translation using the word "of"

$\frac{5}{6} x \cdot \left(\frac{6}{5}\right) = 35 \cdot \left(\frac{6}{5}\right)$

$x = 42$

> $\frac{5}{6}$ of the number 42 is 35

◆**Example 19**   What part of 42 is 28 ?

Let x = the part

x • 42 = 28   Translation using the word "of"

$x \cdot 42 \cdot \left(\dfrac{1}{42}\right) = 28 \cdot \left(\dfrac{1}{42}\right)$

$x = \dfrac{2}{3}$

$\boxed{\dfrac{2}{3} \text{ of 42 is 28.}}$

Name:_____   6.1 Problem Set

For each of the following problems, first approximate your answer, then use your calculator to perform the indicated operation, and finally round your answer to the "least precise" number in the problem.

1) $5.34 + (16.2)(8.4)$ Approximation: Display: Rounded:	2) $(6.4)^2 + (8.1)(3.6)$ Approximation: Display: Rounded:	3) $\sqrt{94} + (18)(3.62)$ Approximation: Display: Rounded:
4) $15.62 \div 4.1 + 3.6$ Approximation: Display: Rounded:	5) $(47.6)(21.48) \div 12.7$ Approximation: Display: Rounded:	6) $\sqrt{29.6} + (13.05)(66.4)$ Approximation: Display: Rounded:
7) $(15.8)(9.725) - (3.2)^2 (7.6)$ Approximation: Display: Rounded:	8) $(15.6)^3 + (8.7)^2 (4.6)$ Approximation: Display: Rounded:	9) $186.78 \div 19.6 - 3.48$ Approximation: Display: Rounded:

Solve the following problems.

10) Find $\frac{2}{3}$ of 84	11) $\frac{3}{4}$ of what number is 18?	12) What part of 35 is 14?
13) What part of 50 is 24?	14) Find $\frac{5}{8}$ of 40	15) Find $1\frac{1}{2}$ of 52
16) $\frac{2}{3}$ of what number is 42?	17) What part of 80 is 12?	18) $\frac{7}{20}$ of what number is 49?

Solve the following equations.

19) $5 + 3(x + 4) = 20$	20) $8 - 4(x + 6) = 18$	21) $\frac{3}{5}x + 4 = 28$
22) $2.4x - 4.8 = 31.2$	23) $1.4x + 4.2 = 22.4$	24) $6 + 0.3x = 1.2$

6.2 Introduction to Percent

We are bombarded daily with invitations to buy things at incredible savings. The advertising headlines promise savings of 40%, 50% and more. Banks flood the media with inducements for us to borrow their money at very reasonable interest rates given in percents. Diet centers beckon us to lose 10% of our weight before Summer swim suit season arrives. We live in a world where people think and discuss percent in many different situations. As consumers, we are expected to understand percent and do quick calculations in our heads. In this section, we will be introduced to percent.

What is Percent?

Percent literally means "per 100" or "out of 100". When one says or writes 8%, the meaning is 8 per 100 or 8 out of 100. The statement, "8% of the population prefer purple cars" does not mean that the population is 100 and 8 of these people prefer purple cars, but rather, that the proportion of people preferring purple cars is equal to 8 out of 100. It is important to see and understand the concept of "out of 100" because this will be the **pattern** used as we work percent problems.

Equivalent Forms of Percent

Since percent means out of 100, we can write any percentage number as a fraction with a denominator of 100.

$$35\% \text{ means 35 out of 100, which means } 35\% = \frac{35}{100}$$

$$\frac{35}{100} = \frac{7}{20}$$

We can say that $35\% = .35 = \frac{7}{20}$. Since these are all equal, we call them **equivalent** forms.

◆ **Example 1:** Write 48% in decimal and common fraction form.

$$48\% = .48 \qquad \text{\% means out of a 100}$$

$$48\% = \frac{48}{100} = \frac{12}{25} \qquad \text{\% means out of a 100, reduce}$$

◆ **Example 2:** Write 0.65 as a percent and as a common fraction.

$$0.65 = 65\% \qquad \text{\% means out of a 100}$$

$$0.65 = \frac{65}{100} = \frac{13}{20} \qquad \text{\% means out of a 100}$$

◆**Example 3**   Write the common fraction $\frac{5}{8}$ as a decimal fraction and as a percent.

$$\frac{5}{8} = 0.625 \qquad \text{Use calculator, divide 5 by 8}$$

$$0.625 = 62.5\% \qquad \text{\% means out of 100}$$

The Percentage Statement

From an arithmetic standpoint, there are basically three types of percentage statements: One where the "**is** number" is missing, one where the "**%** number" is missing and one where the "**of** number" is missing. From an algebraic standpoint there is only one type. We simply write the percentage statement as an equation and solve for the indicated variable.

Arithmetic	Algebra (equation)	Algebra (proportion)
25% of 84 is ____	$(.25)\ 84 = x$	$\dfrac{x}{84} = \dfrac{25}{100}$
____% of 40 is 8	$\dfrac{x}{100}(40) = 8$	$\dfrac{x}{100} = \dfrac{8}{40}$
75% of ____ is 24	$.75\ x = 24$	$\dfrac{75}{100} = \dfrac{24}{x}$

Written in the algebraic form, it does not matter which part of the percent statement is missing. It is just a matter of changing into an equation and solving. Using our mathematical way of thinking, we will be **organizing** each problem involving percent into a percentage statement with one member missing. We use the **pattern** of changing to an equation, and finally **decide** how to use our answer. The combination of all the above, then, becomes the **skill** of solving a percentage type problem.

Helpful Hint 14:   *There are two **patterns** that must be understood in solving percent problems. The first is the meaning of the word "**percent**" and the second is the meaning of the word "**of**" as it is used in a mathematical statement of this kind. As stated previously, think of percent as, "out of a hundred" and think of the word "of" as multiplication when used in a percentage statement.*

The following worked out problems model the process to be used on percent problems.

◆**Example 4**   Solve the percentage statement   _____ % of 72 = 45

$$\frac{x}{100} \cdot 72 = 45 \qquad \text{\% means out of 100, of means multiply}$$

$$\frac{x}{100} \cdot 72 \cdot \left(\frac{100}{72}\right) = 45 \cdot \left(\frac{100}{72}\right) \qquad \text{Solve equation, multiply by reciprocal}$$

$$x = 62.5$$

$$\boxed{62.5\% \text{ of } 72 = 45}$$

♦ **Example 5**   Solve the percentage statement   85% of _____ is 357

$(0.85) \cdot x = 357$ % means out of 100, of means multiply

$x = 420$ Solve for x

> 85% of <u>420</u> is 357

♦ **Example 6**   Solve the percentage statement   28% of 560 is _____

$(0.28) \cdot (560) = x$ % means out of 100, of means multiply

$x = 156.8$ Solve for x

> 28% of 560 is <u>156.8</u>

Solving Percentage Word Problems

To solve a percentage word problem, we will follow the procedure developed and practiced in previous chapters and adapt it to the special circumstances of a percentage problem.

1. **Categories** of information. _____ % of _____ is _____
2. **Numbers** with the categories. Fill in the blanks in the percentage statement above, using two numbers given in the problem and a variable for the unknown quantity. This step involves translation from words to symbols.
3. Form an **equation**. Change percentage statement into an equation or proportion.
4. **Solve** equation or proportion.
5. Write a **sentence** that answers the question.

♦ **Example 7**   During September, Delta Burke lost 8% of her weight. If she began the month at 175 pounds, how much did she weigh at the end of the month?

_____ % of _____ is _____   Categories

<u> 8 </u> % of <u> 175 </u> is _____
% lost   Orig wt   Amt lost   Numbers with categories

$(0.08) \cdot (175) = x$ Form an equation from percentage statement

$x = 14$ Solve

14 pounds is the weight lost, so 175 − 14 = 161

> Delta weighed 161 pounds at the end of the month.

Helpful Hint 15:   *As demonstrated in Example 7, always start a percent word problem with the percentage statement: ___ % of ___ is ___. To ensure that you fill in the correct numbers in the blanks, write the corresponding words from the problem beneath the blanks.*

Chapter 6: Percent

◆**Example 8**   Sally sells sea shells by the sea shore on a commission basis. One week she sold $1580 worth of sea shells and was paid $189.60 as her commission. What was her rate of commission?

$$\underset{\text{\% Comm}}{\underline{\$189.60}} \text{ \% of } \underset{\text{Amt of Sales}}{\underline{\$1580}} \text{ is } \underset{\text{Amt of Comm}}{\underline{}}$$ Categories and numbers

$189.6 = \dfrac{x}{100} \cdot (1580)$ Form an equation

$x = 12$ Solve

| Sally was paid a 12% commission. | Sentence

◆**Example 9**   Laurel and Hardy agreed to paint a 75 year old barn and charge $1200 for their labor. The red paint they used was on sale at the unbelievable price of $19.95 per gallon, plus $8\frac{1}{2}$% sales tax. Laurel took an hour off each day to watch "Days of Our Lives" and only worked 38 hours while poor Hardy painted a full 42 hours. What percent of their total time did Hardy paint and how much of the total $1200 should he get?

Solution: There is much useless information in this problem. The question deals with time and earnings for Hardy so let us gather information about time and earnings.

38 hours + 42 hours = 80 total time

$$\underset{\text{\% time Hardy}}{\underline{}} \text{ \% of } \underset{\text{Total time}}{\underline{80}} \text{ is } \underset{\text{Time Hardy}}{\underline{42}}$$ Categories and numbers

$42 = \dfrac{h}{100} \cdot (80)$ Form an equation

$h = 52.5\%$ Solve

| Hardy painted 52.5% of the total time. |

And be sure to answer the second question. Since Hardy worked 52.5% of the time, he should get 52.5% of the money.

$$\underset{\text{\% due Hardy}}{\underline{52.5\%}} \text{ \% of } \underset{\text{Total pay}}{\underline{\$1200}} \text{ is } \underset{\text{Amt due Hardy}}{\underline{}}$$

$(0.525) \cdot (80) = w$

$W = \$630$

| Hardy should get paid $630 for his labor. |

Name:_____   6.2 Problem Set   229

Write each of the following in decimal fraction form.

1) 45%	2) $8\frac{1}{2}$%	3) 12.7%
4) $2\frac{1}{4}$	5) 15%	6) 125%

Write each of the following in common fraction form, reduced to lowest terms.

7) 35%	8) 6%	9) 140%
10) 37.5%	11) 7.5%	12) $12\frac{1}{2}$%

Write each of the following in percent form.

13) 0.34	14) 0.035	15) 3.09
16) $2\frac{1}{2}$	17) $\frac{3}{8}$	18) 3.9

Estimate the percentage part by shading the figures below.

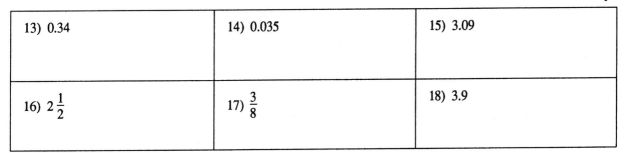

19) Shade 23% of   20) Shade 52% of   21) Shade 84% of

Estimate the percentage "shaded-in" for each of the figures below.

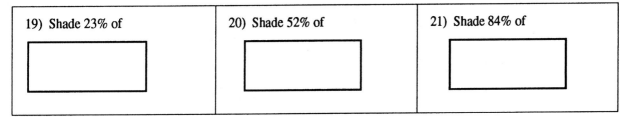

22)   23)   24)

Change each of the following into an equation and solve.

25) 45% of 120 is _____	26) _____ % of 36 is 9	27) 20% of _____ is 8
28) 16% of _____ is 68	29) 60% of 855 is _____	30) _____ % of 420 is 147
31) _____ % of 80 is 16	32) 25% of _____ is 12	33) 15% of 48 is _____
34) 40 is _____ % of 500	35) 3% of 275 is _____	36) 5% of _____ is 21.8

Name:_____

For each of the following, estimate an answer, use a calculator to do the operations and round your answer to the appropriate degree of accuracy.

37) 92.6 + (15.3) • (8.6) Est: Display: Round:	38) (386) • (2.8) + 149.3 Est: Display: Round:	39) (187.50)•(0.5) − (4.62) Est: Display: Round:

Solve the following word problems.

40) Joe lost 8% of his marbles last Saturday. If he started the day with 175 marbles, how many marbles did he end with?	41) Mildred and Mabel panned a total of 5 oz. of gold last summer. If Mildred got $2\frac{2}{5}$ oz., what was her percent of the total?
42) To start school, Olaf spent $90 on books, $18 on paper and $12 on a deluxe Mickey Mouse pencil for his math class. What percent of his total did he spend on paper?	43) Boris earned $140 a week on his part-time job. His boss was impressed with his ability to **organize information,** recognize **patterns** and **make decisions** and granted him a whopping 35% increase in salary. What will his new salary be?

6.3 Practice with Percent

This section is designed to give practice in understanding percent.

Solving a Percentage Statement

The key to solving a percentage statement is to translate the statement into an algebraic equation. The translation will use % to mean "out of 100" and "of" to mean multiply.

◆**Example 1**   Solve the percentage statement.   8.4% of _____ is 303.66

$(.084)x = 303.66$ % means out of 100, of means multiply

$$\frac{0.084}{0.084}x = \frac{303.66}{0.084} \quad \text{Solve}$$

$x = 3615$

> 8.4% of 3615 is 303.66

Solving a word problem

To solve a word problem involving percent, we will follow the procedure developed and practiced in previous section.

1. **Categories** of information. Use empty percent statement. _____ % of _____ is _____
2. **Numbers** with the categories. Fill in the blanks in the percentage statement above, using two numbers given in the problem and a variable for the unknown quantity. This step involves translation from words to symbols.
3. Form an **equation**. Change percentage statement into an equation or proportion.
4. **Solve** equation or proportion.
5. Write a **sentence** that answers the question.

◆**Example 2**   When Kaleb started college, he had the following expenses: $185 for registration, $84.75 for books and $55.25 for supplies. What percent of his total cost was spent on supplies?

First, obtain his total cost: $185 + 84.75 + 55.25 = 325$

_____ % of 325 is 55.25   Translate into a percentage statement

$$\frac{x}{100} \cdot 325 = 55.25$$

$$\frac{x}{100} \cdot 325 \cdot \left(\frac{100}{325}\right) = 55.25 \cdot \left(\frac{100}{325}\right)$$

$x = 17$

1_7_% of 325 is 55.25

> Kaleb spent 17% of his total cost on supplies.

Solve each of the following percent problems.

1) 12 is _____ % of 48	2) 18% of 245 is _____	3) 25% of _____ is 35
4) 120% of 65 is ___	5) 6 is 15% of _____	6) $2\frac{1}{2}$% of 520 is ___
7) 7.5% of 880 is ___	8) 4.5% of _____ is 36	9) _____ % of 448 is 168
10) 40 is ___ % of 480	11) 125% of ___ is 50	12) 7.5% of ___ is 36

6.3 Problem Set

Name: _____

On each of the following problems, estimate the answer, use your calculator and round answer to the appropriate degree of accuracy.

13) $(16.8)(5.06) + (19.6)(2.6)^2$	14) $\sqrt{52.86}\left(\frac{17.4}{7.3}\right) - (3.28)^2$	15) $\frac{(128)(16.4)}{5.26} + \sqrt{79.2}$
Est.	Est.	Est.
Display:	Display:	Display:
Round:	Round:	Round:
16) $\frac{(9.62)^2 \cdot \sqrt{104.94}}{55.2}$	17) $(18.6)(225.4) - 16.2\sqrt{29.4}$	18) $\frac{\sqrt{119.78}}{(1.46)(4.2)} + (3.84)^2$
Est.	Est.	Est.
Display:	Display:	Display:
Round:	Round:	Round:

Solve each of the following word problems.

19) Ernie earned $125 per week and got a 6% raise. What is his new salary?	20) Bonnie had the following expenses last month: $380 for rent, $245 for food, $85 for clothing, $260 for transportation, and $170 for other expenses. What percent of her total did she spend on transportation?

21) Boris bought a new wastebasket at a 28% discount. If the regular price was $72, what was the price that he paid?

22) Arlene paid $2400 for a car and sold it for a 48% loss four years later when she graduated with her math degree and got her first real job. For how much did she sell her car?

23) Ferd Burfil was excited to get a 20% discount and pay only $12 for a cap with a propeller on top. What was the regular price of his wonderful cap?

24) Roseanne Barr bought a 5 pound box of candy to supplement her new diet. This candy costs $5.60 per pound and the sales tax was $8\frac{1}{2}$%. When she got her change from giving the clerk $40, she decided to spend the remainder buying some potato chips. How much did she have left to purchase the potato chips?

6.4 Practice with Percent

This section is designed to give additional practice in understanding percent.

Solving a Percentage Statement

The key to solving a percentage statement is to translate the statement into an algebraic equation. The translation will use "%" to mean "out of 100" and "of" to mean multiply.

◆**Example 1**   Solve the percentage statement.   ____ % of 744 is 279

$$\frac{x}{100} \cdot 744 = 279$$

$$\frac{x}{100} \cdot 744 \cdot \left(\frac{100}{744}\right) = 279 \cdot \left(\frac{100}{744}\right)$$

$$x = 37.5$$

$\boxed{37.5\% \text{ of } 744 \text{ is } 279}$

Solving a Word Problem

To solve a word problem involving percent, we will follow the procedure developed and practiced in the previous section.

1. **Categories** of information. Use an empty percent statement: ____% of __ is ___.
2. **Numbers** with the categories. Fill in the blanks in the percentage statement above, using two numbers given in the problem and a variable for the unknown quantity. This step involves translation from words to symbols.
3. Form an **equation**. Change percentage statement into an equation or proportion.
4. **Solve** equation or proportion.
5. Write a **sentence** that answers the question.

◆**Example 2**   Olga rebuilt a car and sold it for a 35% profit on her cost. Her costs were $2465 to buy the car and $4035 to fix it up. What was the selling price of the car?

First we must obtain her total cost: $2465 + $4035 = $6500

35% of $6500 is _____ (the profit)   Percent statement, fill in with translation

$(0.35) \cdot (\$6500) = x$

$x = \$2275$ the profit.

Cost + Profit = Selling Price

$6500 + $2275 = Selling Price

$\boxed{\text{The selling price of the car was \$8775.}}$

Find the missing value in each of these percent problems.

1) 2.6% of 425 is _____

2) 150 is _____ % of 60

3) 60 is _____ % of 150

4) 55% of _____ is 1529

5) $4\frac{1}{2}$ % of _____ is 387

6) 145% of 3226 is _____

7) 18 is _____ % of 360

8) $2\frac{1}{4}$ % of 6280 is _____

9) 175% of _____ is 28

10) 37.5% of 488 is _____

11) 132 is _____ % of 5280

12) 12.5% of _____ is 608

Solve the following problems.

13) Olaf sold olives on commission. One week he sold $3680 worth of olives and was paid $294.40 in commission. What was his rate of commission?

14) Fannie Farkel found 15 faulty faucets at the Faust Faucet Factory. If this represents $1\frac{1}{4}\%$ of the total production for a day, how many faucets were produced that day?

15) Batman paid $38,750 to have his Batmobile made. He later sold it to a collector who paid $42,500. What was the percent gain on the sale?

16) Oliver was concerned about his 240 pound weight and went on a diet, losing 10% of his weight the first month. The second month, he gained 10%. How much did he weigh at the end of the second month?

17) Ivan scored 95% when he answered 38 questions correctly on a multiple choice history test. If all the questions were of equal value, how many questions were on the test?

18) The cost of repairing Frank's washing machine was $158. If the parts cost $60.04 and the rest was for labor, what percent of the total cost was labor?

6.5 Chapter Review

The goals of this chapter include:

1. Approximating and rounding answers in arithmetic calculations.
2. Using a hand held calculator in arithmetic calculations.
3. Solving percent problems.
4. Solving word problems involving percent.

Using a Calculator   [6.1]

In this chapter you have done many calculations using a hand held calculator. In each case you are expected to approximate in order to determine if the calculator display has a reasonable answer. To approximate calculations, use "easy to use" numbers rather than the numbers given in the problem. In some cases you are expected to round-off your answer.

◆**Example 1** Use your calculator to simplify. $\sqrt{(6.8)(9.2)} + \dfrac{14.72}{2.94}$

$= \sqrt{7 \cdot 9} + \dfrac{15}{3}$ Change to "easy to use" numbers

$= \sqrt{63} + 5$ Calculate with "easy" numbers

$= \sqrt{64} + 5$ Change to "easy to use" numbers

$= 8 + 5$

$= \boxed{13}$ Approximate answer

$\sqrt{(6.8)(9.2)} + \dfrac{14.72}{2.94} = 12.916291$ Calculator display

$\boxed{12.9}$ Round to least precise (nearest tenths)

Solving Percent Problems   [6.2]

To solve a percentage statement, first change to an equation or proportion using a variable to represent the unknown part and then solve for the variable. In translating the statement, think of percent as "out of 100" and the word "of" means multiplication.

◆**Example 2** Solve: 12.5% of _____ is 10.75

$0.125 x = 10.75$ % means out of 100, of means multiply

$\dfrac{0.125x}{0.125} = \dfrac{10.75}{0.125}$ Solve for x

$x = 86$ so $\boxed{12.5\% \text{ of } \underline{86} \text{ is } 10.75}$

Solving Word Problems Using Percent   [6.3]

To solve a word problem involving percent, follow the procedure developed and practiced in this chapter.

1. **Categories** of information. Use empty percent statement: _____ % of _____ is _____
2. **Numbers** with the categories. Fill in the blanks in the percentage statement above, using two numbers given in the problem and a variable for the unknown quantity. This step involves translation from words to symbols.
3. Form an **equation**. Change percentage statement into an equation or proportion.
4. **Solve** equation or proportion.
5. Write a **sentence** that answers the question.

◆**Example 3**   Wilma walked 25% more than Waldo each week. If Waldo walked a total of 28 miles each week, how many miles did Wilma walk?

25% of 28 is x	**Categories** and **numbers**.
$(0.25) \cdot (28) = x$	Form an **equation**
$7 = x$	**Solve**
$28 + 7 = 35$	
Wilma walked a total of 35 miles each week	**Sentence**

Name:_____   6.5 Chapter Review Test   245

Solve each of the following problems.

1) 75% of 1024 is _____	2) _____ % of 72 is 27	3) $2\frac{1}{2}$% of _____ is 16.35
4) 28% of 430 is _____	5) $2\frac{1}{4}$% of _____ is $4\frac{1}{2}$	6) _____% of 550 is 99
7) _____ % of 560 is 36.4	8) 8% of _____ is 35.2	9) $12\frac{1}{2}$% of 104 is _____

On each of the following, approximate an answer, use a calculator and show the display, and finally give the answer rounded to the required place of accuracy.

10) $18.6 + (2.3)(879.6)$ Est: Display: Round:	11) $\dfrac{416.2}{5.86} - (5.9)$ Est: Display: Round:	12) $\dfrac{(13.4) \cdot (6.84)}{15.4} + \dfrac{826.1}{20.2}$ Est: Display: Round:

Solve the following word problems.

13) A family spent 62% of its budget on essentials, 26% on frivolous items and saved the rest. If their monthly income was $2700, how much money did they save each month?	14) Filbert was paid $720 per month in salary plus 6% commission on all the almonds that he sold. Last year he sold a total of $286,400 worth of almonds. What was his total earnings for the year?

15) Javier sold computer software and was paid 4% for the first $50,000 in sales and 7% for sales over the $50,000 goal. Find his total commission for the month that he sold $74,000 worth of software.

16) Barbara bought furniture with a regular price of $4800. She got a 30% discount on the first $2500 and 40% discount on the amount which exceeded $2500. How much was she charged for the furniture?

17) Laurel and Hardy were hired to move 40 pianos. If they moved only 14 pianos in the morning and the rest in the afternoon, what percent of the pianos were moved in the afternoon?

18) Olaf bought seven CD's from his friend Boris who gave him a 25% discount. If the regular price of each CD was $12, what was the total price Olaf paid including the 8.5% sales tax?

19) Tarzan made a salad consisting of four pounds of lettuce, three pounds of tomatoes, two pounds of squash, one-half pound of garlic and $2\frac{1}{2}$ pounds of celery. What percent of the salad weight was the lettuce?

20) Find two numbers whose sum is 8 such that five times the smaller increased by twice the larger is 23.

21) The larger of two numbers is one more than three times the smaller. When four is multiplied by the larger and diminished by five times the smaller, the result is 18. Find the larger number.

22) Boris bought big baskets at a 45% discount. The regular price was $65 and he paid with a crisp $50 bill. If he spent the remainder buying a pizza for himself and Olga, how much did the pizza cost?

CHAPTER 7
Graphs

7.1 Introduction to Graphs

In this chapter, you will learn to:

1. *Interpret graphs.*
2. *Graph a linear equation.*

In this chapter we will look at some graphs that are already made and interpret the information contained in the graph. We will also learn how to make a graph that represents a linear equation.

Using Graphs

One of the primary ways that a mathematician organizes information and looks for patterns is with graphs. Looking at a daily newspaper, we can see many examples of information given by graphs. A graph is a means of visualizing a relationship that exists between sets of numbers. Because of the prevalence of graphs in our daily lives, it is important that we read and understand the information given in a graph. The graph below describes the cost of tuition at the University of California over the years between 1983 and 1993. We are looking at the relationship that exists between the set of numbers representing years and the set of numbers representing the cost of tuition at the University of California.

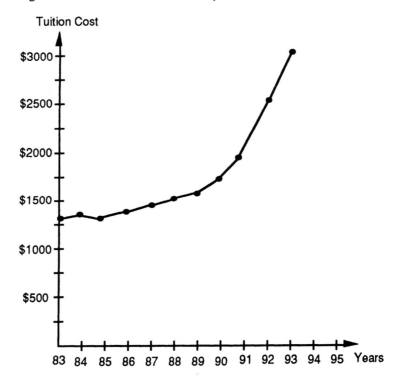

By looking at the graph, we can see the steep increase in cost over the past few years. The graph can be used to find approximate answers to many questions such as:

1, What was the tuition in 1984?

    Answer: Approximately $1300

2. Has the tuition ever gone down in the past few years? If so, when and by how much?

   Answer: Yes, during the school year 1984-85 the tuition declined by about $50.

3. In what year was the tuition about $1500

   Answer: During the 1988-89 school year.

4. Without performing calculations, what was the approximate yearly tuition from 1983-87?

   Answer: Approximately $1400.

5. During which years were the greatest tuition increases?

   Answer: During the years 1991-92 and 1992-93.

6. In what year was the tuition about $2500?

   Answer: During the 1992-93 school year.

7. During which years was the tuition below $2000?

   Answer: During the years 1983 to 1991.

8. In what year was the tuition about $1800?

   Answer: During the 1991-92 school year.

Making Graphs

In order to understand how a graph is made, let us look at some points in a coordinate system. The location of a point is given by using an ordered number pair, written inside a parenthesis. The first number in the parenthesis gives the location in the **horizontal** direction, the second number gives the **vertical** direction. The horizontal axis is normally called the x-axis, the vertical axis is normally called the y-axis. The point described by the ordered pair (2,3) is the point located 2 units to the right of the **origin** (the point where the x-axis and the y-axis meet) and 3 units up from the origin.. Note the ordered pairs assigned to the points on the graph :

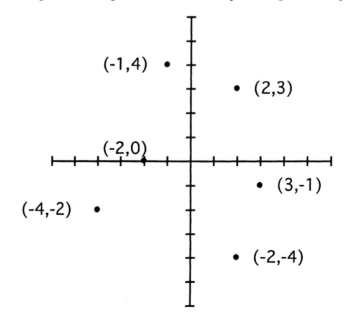

Consider the points of the graph below. Do you see any **pattern**?

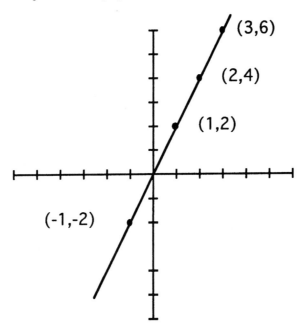

Some of the ordered pairs are: $(-1,-2)$ $(1,2)$ $(2,4)$ and $(3,6)$

Can you see that in each case the y value (2nd number) is twice the size of the x value (1st number). The pattern could be stated as:

$$y = 2x$$

Graph Language

In graph language, the y value is called the **ordinate** and the x value is called the **abscissa**. With this language we can describe the pattern of how the y value relates to the x value.

Helpful Hint 16: *Always place the variable representing the abscissa to the immediate right of the = sign and place the constant, if any, to the right of the abscissa. This form is often written as y = ax + b with y = the ordinate, the x term = the abscissa and b = the constant.*

◆**Example 1** Translate into an equation: The ordinate is two more than the abscissa.

  Translation: $y = x + 2$

◆**Example 2** Translate: The ordinate is four more than three times the abscissa.

  Translation: $y = 3x + 4$

◆**Example 3** Translate: The ordinate is three less than twice the abscissa.

  Translation: $y = 2x - 3$

◆**Example 4** Translate: The ordinate is five more than one-half the abscissa.

  Translation: $y = \frac{1}{2} x + 5$

◆**Example 5** Translate: The ordinate is the square root of the sum of the abscissa and two.

  Translation: $y = \sqrt{x + 2}$

◆**Example 6** Translate: The ordinate is two more than the square root of the abscissa.

  Translation: $y = \sqrt{x} + 2$

◆**Example 7** Translate and graph: The ordinate is three less than twice the abscissa.

  Translation: $y = 2x - 3$

  Graph by choosing a few representative points. We do this by selecting a few values for x (the abscissa) and finding the corresponding y values (the ordinate). For convenience, let us choose x-values of $\{-1, 0, 1, 2\}$

  Choose $x = -1$ $y = 2(-1) - 3$ or $y = -5$ and the point is $(-1, -5)$

  Choose $x = 0$ $y = 2(0) - 3$ or $y = -3$ and the point is $(0, -3)$

  Choose $x = 1$ $y = 2(1) - 3$ or $y = -1$ and the point is $(1, -1)$

  Choose $x = 2$ $y = 2(2) - 3$ or $y = 1$ and the point is $(2, 1)$

We can now graph these points and connect them with a straight line as seen in the graph below.

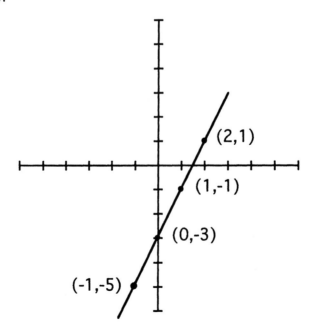

◆**Example 8** Translate and graph using the given x-values:

Translate and graph using the given x-values: The ordinate is two more than negative three times the abscissa. $x = -2, x = -1, x = 0, x = 1$

$y = -3x + 2$ Translation

If $x = -2$ then $y = -3(-2) + 2$

$y = 6 + 2$ or $y = 8$ so the point is **(-2, 8)**

If $x = -1$ then $y = -3(-1) + 2$

$y = 3 + 2$ or $y = 5$ so the point is **(-1, 5)**

If $x = 0$ then $y = -3(0) + 2$

$y = 0 + 2$ or $y = 2$ so the point is **(0, 2)**

If $x = 1$ then $y = -3(1) + 2$

$y = -3 + 2$ or $y = -1$ so the point is **(1, -1)**

These points then graph as pictured below and when connected, form a straight line.

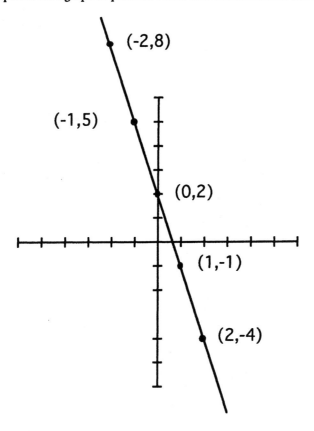

1) Use the graph below to locate each of the following points.

　　A (2, 4)　　B (−3, 1)　　C (−2, −1)　　D (4, 0)
　　E (−2, 0)　　F (3, −2)　　G (0, −1)　　H (−5, −2)

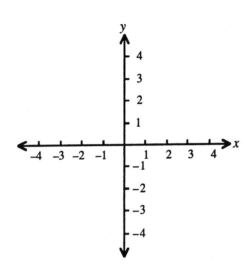

2) Use the graph below to name the coordinate of the points on the graph.

　　A (,)　　B (,)　　C (,)　　D (,)
　　E (,)　　F (,)　　G (,)　　H (,)

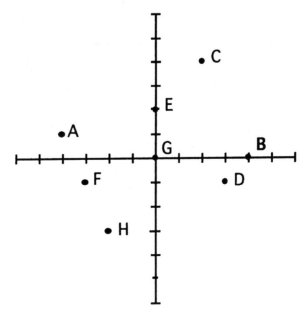

7.1 Problem Set

Translate each of the following.

3) The ordinate is six more than five times the abscissa.

___

4) The ordinate is three more than four times the abscissa.

___

5) The ordinate is two less than three times the abscissa.

___

6) The ordinate is four less than one-half the abscissa.

___

7) The ordinate is six more than two-thirds the abscissa.

___

8) The ordinate is five more than twice the square of the abscissa.

___

Translate and find the indicated missing values.

9) The ordinate is three more than five times the abscissa.

(0,   )   (4,   )   (−6,   )   (−3,   )

___

10) The ordinate is five more than three times the abscissa.

(−2,   )   (− 1,   )   (0,   )   (4,   )

Translate and graph the points indicated. Connect the points with a line.

11) The ordinate is one more than three times the abscissa.

（ –3,   ）   (–2,   )   (0,   )   (2,   )   (3,   )

12) The ordinate is three less than twice the abscissa.

(–2,   )   (–1,   )   (0,   )   (1,   )   (2,   )

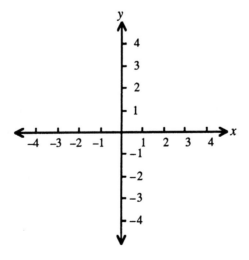

13) Ferd Burfil did a gas efficiency test on his car and graphed the results below:

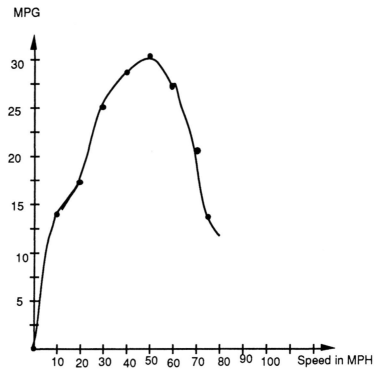

a) What was the gas mileage at 20 MPH?

b) At what speed was the gas mileage about 15 MPG?

c) At what speed was the gas mileage the best?

d) What was the MPG when the speed was 30 MPH?

e) What was the MPG when the speed was 80 MPH ? Why was it low at this speed?

f) What was the speed when the gas mileage was above 25 MPG?

7.2 Making Graphs

In this section we will continue to develop the skill of translating from words into mathematical symbols and use the translation to represent an equation in two variables as a graph in the coordinate plane.

Translations of Words into an Equation

When we graph an equation, we are plotting points on a coordinate plane. The coordinate plane is made up of a vertical direction and a horizontal direction. The horizontal component is called the **abscissa** (usually the x-value) and the vertical component (usually the y-value) is called the **ordinate**.

◆**Example 1** Translate: The ordinate is six more than five times the abscissa.

$$y = 5x + 6$$

◆**Example 2** Translate: The ordinate is five less than twice the abscissa.

$$y = 2x - 5$$

◆**Example 3** Translate: The ordinate is four more than two-thirds of the abscissa.

$$y = \frac{2}{3}x + 4$$

◆**Example 4** Translate: The ordinate is seven less than twice the square of the abscissa.

$$y = 2x^2 - 7$$

◆**Example 5** Translate: The ordinate is nine more than the square root of three times the abscissa

$$y = \sqrt{3x} + 9$$

Drawing a Graph

To draw the graph of an equation in two variables we choose values of the x-variable and plug these values into the equation to find the corresponding y-values. In most cases the choice of which values of x to use is arbitrary. That is we can choose any value of x we want and then find out the value of y that would result when we "plug-in" the x-value. Since the choice of x-values is arbitrary, we tend to use easy to plug-in values for x.

◆Example 6 Translate and graph: The ordinate is five less than twice the abscissa.

$$y = 2x - 5 \qquad \text{Translation}$$

As arbitrary choices, let us choose $x = -1, x = 0, x = 1, x = 2, x = 3$

If $x = -1$ then, $y = 2(-1) - 5$

$y = -7$ so one point of the graph is $(-1,-7)$

If $x = 0$, then, $y = 2(0) - 5$

$y = -5$ so a second point of the graph is $(0,-5)$

If $x = 1$ then $y = 2(1) - 5$

$y = -3$ so a third point of the graph is $(1,-3)$

If $x = 2$, then $y = 2(2) - 5$

$y = -1$ so a fourth point of the graph is $(2,-1)$

If $x = 3$, then $y = 2(3) - 5$

$y = 1$ so a fifth point of the graph is $(3,1)$

The points are graphed below and connected with a straight line, hence the name, "linear equation".

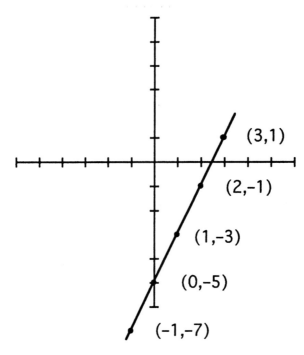

Translate each of the following.

1) The ordinate is eight more than negative five times the abscissa.

2) The ordinate is four less than seven times the abscissa.

3) The ordinate is six more than three-fifths the abscissa.

4) The ordinate is seven less than five-eighths the abscissa.

5) The ordinate is three more than twice the sum of the abscissa and one.

6) The ordinate is five more than the square of the abscissa.

7) The ordinate is five more than the square root of the abscissa.

8) The ordinate is six less than four times the square of the abscissa.

9) The ordinate is two more than the square root of the difference of three times the abscissa and four.

10) The ordinate is three less than twice the square of the abscissa.

11) The ordinate is the same as the abscissa.

12) The ordinate is five.

13) The abscissa is three.

262   *7.2 Problem Set*   Name:_____

Translate each of the following, find points having the given abscissa and graph. Connect the points to form a line.

14) The ordinate is four more than negative three times the abscissa.

    (–2,   )   (–1,   )   (0,   )   (1,   )   (2,   )

15) The ordinate is two less than three-fourths the abscissa.

    (–8,   )   (–4,   )   (0,   )   (1,   )

16) The ordinate is one more than two-thirds the abscissa.

    (–6,   )   (–3,   )   (0,   )   (3,   )

17) The ordinate is the square root of the difference of the abscissa and three. (This is **NOT** a straight line.)

    (3,   )   (4,   )   (7,   )   (12,   )

18) The ordinate is three less than the square of the abscissa. (This is **NOT** a straight line)

(−3,)   (−2,)   (−1,)   (0,)   (1,)   (2,)   (3,)

19) The ordinate is the opposite of the abscissa.

(−3,)   (0,)   (2,)   (5,)

The graph below shows the hourly sales at Fannie's Famous Funkie Footwear Store.

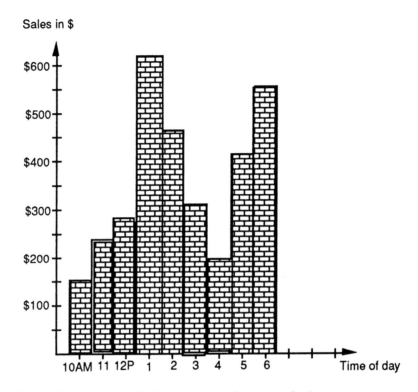

Note: The time of day indicates the end of an hour of sales.

20) What time of day had the greatest amount of sales, and how much was sold during that hour?

21) What time of day had the smallest amount of sales and how much was sold during that hour?

22) What times of day did the sales exceed $500?

23) Give some reasons that might justify why sales were high at certain times of the day.

24) How would you use the above data to decide when to hire part-time help?

7.3 The Function Notation

The function notation is often used in graphing. In this section we will do graphing using this useful notation.

What is the Function Notation?

The function notation is another way of writing the ordinate. The function notation, written as f(x) (or g(x) or h(x) or k(x),etc) is a way of expressing the relationship between two sets of numbers. The notation, f(x) is read "f of x" not f times x. Whenever the function notation appears think of it as the same thing as the ordinate. That is f(x) means the same as y; g (x) means the same as y. One can choose any letter to represent the function so the choice of using g(x) or f(x) or h(x) is up to the one using the function. The function notation is a handy way to express a dependence relation. For example if you earned $8 and hour, your weekly wage would be a function of how many hours you worked and could be expressed using the function notation: W(x) = 8x. You could calculate your wage, W, by substituting the number of hours worked for the variable x.

◆ **Example 1** Translation using the function notation: The ordinate is one more than twice the abscissa.

   Translation: f(x) = 2x + 1

◆ **Example 2** Translation using the function notation: The ordinate is three less than five times the abscissa.

   Translation: g(x) = 5x − 3

◆ **Example 3** Translate with function notation: The ordinate is five more than the square of the abscissa.

   Translation: h(x) = x + 5

◆ **Example 4** Translate with function notation. The ordinate is six more than the square root of the sum of the abscissa and two.

   Translation: k(x) = 6 +

Function Notation used in Word Problems

◆**Example 5** You have a job that pays you $12 per day plus $6 for each hour worked. It can be said that your earnings are a function of time and therefore can be written in the function notation. Express your pay as a function of time and calculate your pay if you worked 25 hours.

Let x = the number of hours worked

Then p(x) = 6x + 12

p(25) = 6(25) + 12

p(25) = 162

Your pay would be $162 for 25 hours work.

◆**Example 6** A TV game show pays each of its contestants $200 plus $75 for each correct answer given during the program. Express a contestant's winnings as a function of number of questions answered. Using the function notation, find how much money is earned by contestant who answered nine questions right.

Let x = the number of correct answers

Then W(x) = 75x + 200

if x = 9 then W(9) = 75(9) + 200

= 675 + 200

= 875

The contestant would win $875 by answering 9 questions correctly.

The Function Notation used in Graphing

The function notation is particularly useful in graphing because once the function is written we can simply replace the variable with "easy-to-use" numbers to find a few representative points, and then graph these points.

◆**Example 7** Translate and graph using the given values of x: The ordinate is two less than three times the abscissa. Sample values of x are: $x = \{-2, -1, 0, 2, 3\}$

Translation: $f(x) = 3x - 2$

Evaluate when $x = -2$, $f(-2) = 3(-2) - 2$

$f(-2) = -8$ so the point **(–2, –8)** is on the graph.

Evaluate when $x = -1$, $f(-1) = 3(-1) - 2$

$f(-1) = -5$ so the point **(–1,–5)** is on the graph.

Evaluate when $x = 0$, $f(0) = 3(0) - 2$

$f(0) = -2$ so the point **(0,–2)** is on the graph.

Evaluate when $x = 2$ $f(2) = 3(2) - 2$

$f(2) = 4$ so the point **(2,4)** is on the graph.

Evaluate when $x = 3$ $f(3) = 3(3) - 2$

$f(3) = 7$ so the point **(3,7)** is on the graph.

These points are graphed below and connected by a straight line. This is called a **linear** function because its graph is a **line**.

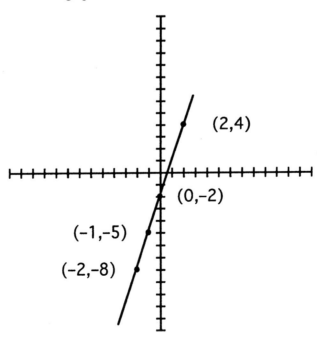

Chapter 7: Graphs

◆**Example 8** Translate and graph using the sample values of x: The ordinate is five less than two-thirds the abscissa. Sample values of x are: $x = \{-3, 0, 3, 6, 9\}$

Translation: $\qquad g(x) = x - 5$

    Evaluate when $x = -3$ $g(-3) = -5$

$g(-3) = -2 - 5 = -7$ So the point $(-3, -7)$ is on the graph

    Evaluate when $x = 0$ $g(0) = -5$

$g(0) = 0 - 5 = -5$ So the point $(0, -5)$ is on the graph

    Evaluate when $x = 3$ $g(3) = -5$

$g(3) = 2 - 5 = -3$ So the point $(3, -3)$ is on the graph

    Evaluate when $x = 6$ $g(6) = -5$

$g(6) = 4 - 5 = -1$ So the point $(6, -1)$ is on the graph

We now have enough sample points to graph the function.

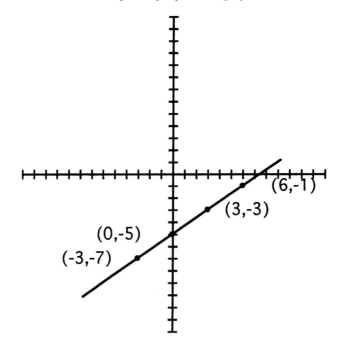

Translate each of the following using the function notation.

1) The ordinate is five less than seven times the abscissa.

2) The ordinate is eight less than twice the abscissa.

3) The ordinate is six more than four times the abscissa.

4) The ordinate is nine more than six times the abscissa.

5) The ordinate is three less than half the abscissa.

6) The ordinate is one more than the square of the abscissa.

7) The ordinate is two less than five times the square of the abscissa;

8) The ordinate is four less than the square root of the abscissa.

9) The ordinate is four less than six times the square of the abscissa.

10) The ordinate is nine more than three times the square root of the difference of the abscissa and two.

11) The ordinate is five less than eight times the square root of the sum of twice the abscissa and one.

7.3 Problem Set

Translate using function notation, find values of the function using the given values, and graph.

12) Boris is paid a wage of $6 a day plus $5 for each hour worked that day. Let x represent the number of hours worked and write his daily wage as a function of the number of hours worked. Use this function to calculate his wage for the sample number of hours worked.

$W(x) = $

$W(2) = $ $W(5) = $ $W(7) = $ $W(8) = $

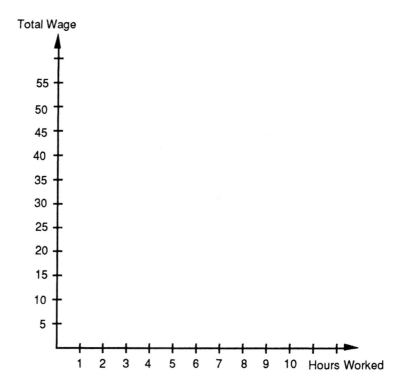

Using the above graph, find each of the following. (DO NOT CALCULATE)

13) Find W(1)

14) Find W(4)

15) Find x if W(x) = 21

16) If W(x) = 36, how many hours did Boris work?

17) Olga is paid $15 per day plus a commission of one-eighth of all her sales. Write a function to describe her pay, use the sample sales figures with x representing her sales and graph the function.

$$P(x) =$$

P(16) =   P(40) =   P(56) =   p(96) =

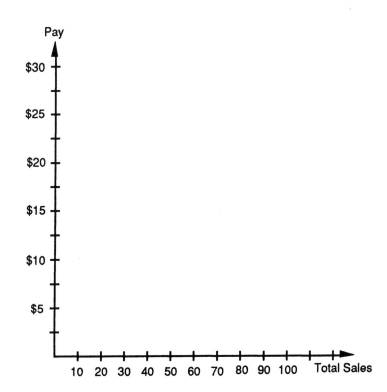

Use the above graph to answer the following questions: (DO NOT CALCULATE)

18) Find P(8) =

19) Find P(72) =

20) If P(x) = 19 find the value of x.

21) If P(x) = 26 find the total amount of Olga's sales.

7.4 Chapter Review

The goals of this chapter include:

1. Interpret graphs.
2. Graph a linear equation.

Interpreting a Graph   [7.1]

A graph is a visual way of organizing information. By learning how to read graphs, we can obtain information quickly and make comparisons easily. We must be able to use the data on the horizontal axis as well as the vertical axis.

◆ **Example 1**   The following graph represents the yearly earnings of a recent college graduate relative to his age. Study the data and answer the questions below the graph.

1) What was the yearly salary at age 23?

   Answer: The salary at age 23 was about $17,000 per year.

2) At what age was the salary about $20,000 per year?

   Answer: At age 25 and again at age 29 the salary was about $20,000 per year.

3) Why did the salary drop at age 28?

   Answer: The graph cannot give an explanation for this kind of event. This person may have been laid-off from his job or a down turn in the economy may have caused a reduction in pay. It is impossible to know.

4) What was the amount of his salary loss at age 28?

Answer: He lost about $12,000 as compared to the previous year.

5) What was the amount of his salary gain going from age 30 to age 31?

Answer: He gained about $10,000 as he went from age 30 to age 31.

Making Graphs   [7.1, 7.2, 7.3]

In this chapter we have mainly made graphs by translating word into an equation and then evaluating the resulting equation by using representative x-values. The x-value, called the **abscissa,** is the horizontal component of the point. The y-value called the **ordinate** is the vertical component. A useful way of representing the ordinate is to use the function notation, $f(x) = y$

◆**Example 2**   Translate and graph using the given x-values. The ordinate is four more than three times the abscissa. Let $x = \{-3, -2, -1, 0\}$

Translation: $f(x) = 3x + 4$

If $x = -3$, then $f(-3) = 3(-3) + 4 = -5$, so the point $(-3,-5)$ is a point of the graph.

If $x = -2$, then $f(-2) = 3(-2) + 4 = -2$, so the point $(-2,-2)$ is a point of the graph.

If $x = -1$, then $f(-1) = 3(-1) + 4 = 1$ so the point $(-1,1)$ is a point of the graph.

If $x = 0$, then $f(0) = 3(0) + 4 = 4$, so the point $(0,4)$ is a point of the graph.

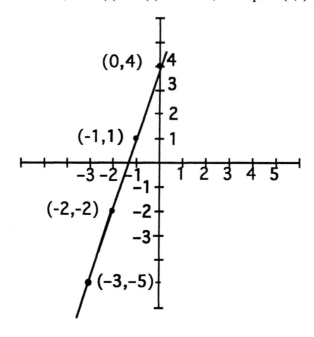

Name:_____   7.4 Chapter Review Test

Translate each of the following sentences using function notation and find the indicated value of the function.

1) The ordinate is eight more than three times the abscissa.

$$f(x) =$$

$f(-5) =$

$f(3) =$

$f(0) =$

---

2) The ordinate is five less than six times the abscissa.

$$g(x) =$$

$g(-4) =$

$g(3)$

$g(7) =$

---

3) The ordinate is one more than two-thirds the abscissa.

$$h(x) =$$

$h(-6) =$

$h(3) =$

$h(6) =$

278   7.4 Chapter Review Test   Name:_____

Find the indicated value of the function and graph. Connect the points to form a line.

4) The ordinate is six less than three times the abscissa.

f (x) =

f (–1) =   f (0) =   f(2) =   f (3) =

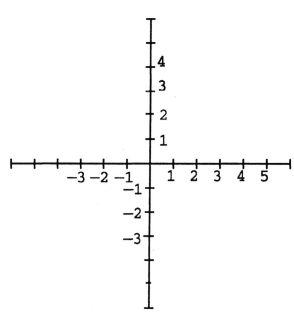

5) The ordinate is five more than negative two times the abscissa.

g(x) =

g(–1) =   g (0) =   g (2) =   g (4) =

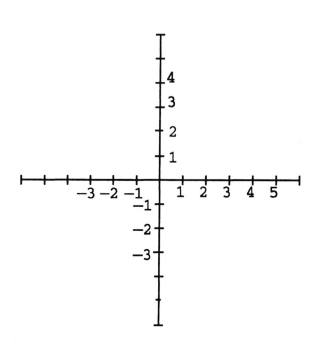

Name:_____   7.4 Chapter Review Test   279

6) The ordinate is the square root of the sum of the abscissa and three.

h (x) =

h (–3) =   h (–2) =   h (1) =   h (6) =

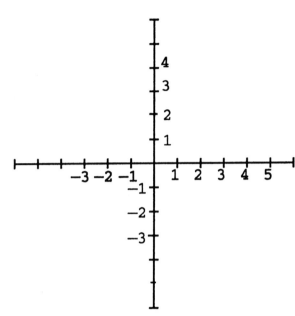

7) The ordinate is three more than twice the abscissa.

f (x) =

f (3) =   f (–2) =   f (0) =   f (1) =

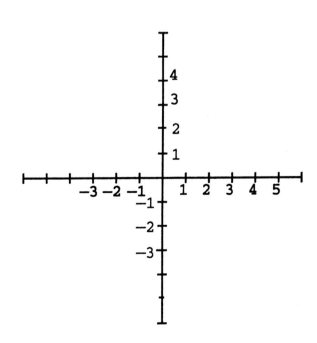

Use the graph, f(x) graphed below, to answer the following questions.

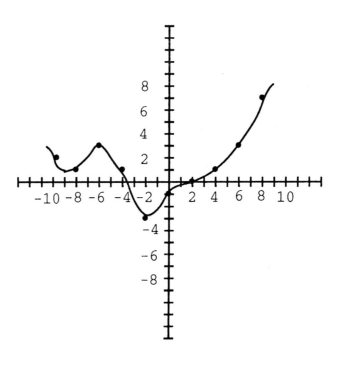

7) What is f(−3)?

8) What is f(0)?

9) For what values of x, if any, does f(x) = 4?

10) For what values of x, if any, does f(x) = 0?

11) What is f(4)?

12) What is f(6)?

13) For what values of x, if any, is f(x) greater than 4?

14) For what value of x is f(x) the greatest? What is the greatest value of f(x)?

15) For what value of x is f(x) the least? What is the least value of f(x)?

CHAPTER 8
Geometry

In this chapter, you will learn to:

1. *Use geometric formulas to solve problems involving rectangles, triangles and circles.*
2. *Apply the Pythagorean theorem to solve right triangle problems.*
3. *Use a combination of formulas to solve irregular figures.*
4. *Use dimensional analysis to convert units of measure.*

8.1 Using Formulas from Geometry: Rectangle and Triangle

In this section we will discuss the geometric figures called polygons.

What Is a Polygon ?

A **polygon** is a closed plane figure determined by three or more line segments. The line segments are called sides of the figure. There are many types of polygons but we will deal only with two of the most common types: rectangles and triangles. We will solve problems involving perimeters and areas. **Perimeter** means the distance measured around a polygon while **area** means the surface measured inside the polygon. Examples of polygons having 3 sides (a triangle), 4 sides (a quadrilateral) and 5 sides (a pentagon)are shown below:

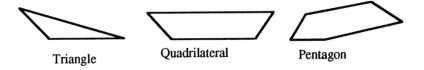

Triangle　　　Quadrilateral　　　Pentagon

What Is a Rectangle ?

A **rectangle** is a four sided polygon whose opposite sides are equal and each of whose angles is a **right angle** (90°). Note that a **square** is a rectangle with all 4 sides equal. Examples of rectangles are shown below:

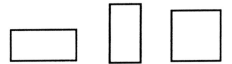

Perimeter of a Rectangle

Since perimeter means distance around a polygon, we can find the perimeter of a rectangle by adding the lengths of the 4 sides. If we let "l" represent the longer side and "w" represent the shorter side, we can express the sum of the four sides with the formula, P = l + w + l + w. Because the opposite sides of a rectangle are equal, we can simplify by adding the equal sides and write this formula as P = 2l + 2w.

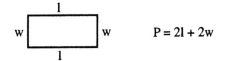

Area of a Rectangle

The area of a rectangle is the surface inside its borders (sides) and is measured in square units. Area can be used to describe the size of the floor of a room, size of a chalk board, or the size of a parking lot. For example a rectangle measuring 3 feet by 2 feet would have 6 square feet as its area. We will write 6 square feet as 6 ft^2.

A rectangle measuring 4 meters by 3 meters would have an area of $12m^2$.

The pattern indicates that we multiply the lengths of the two sides to obtain its area. To do this, we use the formula A = lw.

Organization of a Geometry Problem

As mathematicians, we need an organizational approach to solve problems. Use the following steps to solve geometric problems.

1. **Draw** and **label** the figure to represent the problem.
2. Write the **formula** to be used.
3. **Substitute** given numbers into the formula. (Remember empty parenthesis)
4. **Perform** the given operations or **solve** the resulting equation.
5. Write a **sentence** to answer the question.

◆**Example 1** Find the area and perimeter of a rectangle whose length is 12 feet and whose width is 5 feet.

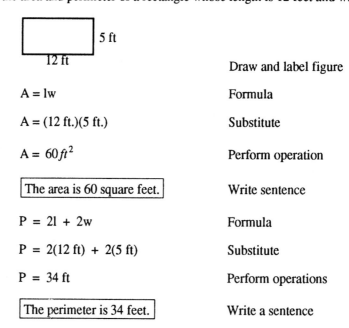

◆**Example 2** Find the width of the rectangle whose length is 52 meters and whose perimeter is 138 meters.

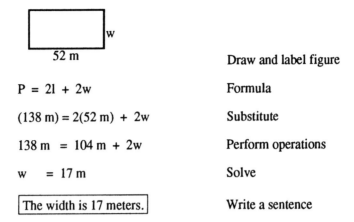

P = 2l + 2w	Formula
(138 m) = 2(52 m) + 2w	Substitute
138 m = 104 m + 2w	Perform operations
w = 17 m	Solve
The width is 17 meters.	Write a sentence

What Is a Triangle ?

A **triangle** is a polygon with 3 sides. The point where two sides meet is called a **vertex**. Each triangle has 3 vertices. The **height** of a triangle is the measurement of the perpendicular segment drawn from a vertex to the line containing the opposite side. Whichever side is chosen to measure the height is called the **base** of the triangle. Consider the three triangles below labeled △ ABC which show how the height measurement is made. The base in each triangle below is the segment AC.

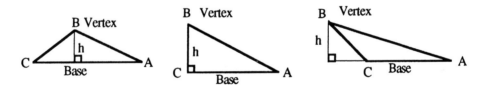

The Area of a Triangle

The area of a triangle is given by the formula,

$$A = \frac{1}{2}bh$$

with b representing the base and h representing the height of the triangle.

The Perimeter of a Triangle

The perimeter of a triangle is the distance around the triangle. The pattern is the addition of the three sides. With a, b, and c representing the sides, we have the formula,

P = a + b + c

Chapter 8: Geometry

◆**Example 3** Find the area and perimeter of the triangle with sides of 8 feet, 9 feet and a base of 15 feet. The height measured from the base is 4 feet.

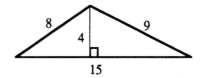

P = a + b + c	Formula
P = (8 ft) + (9 ft) + (15 ft)	Substitute
P = 32 feet	Perform given operations
The perimeter is 32 feet.	Write a sentence
$A = \frac{1}{2}bh$	Formula
$A = \frac{1}{2}(15\text{ft})(4\text{ft})$	Substitute
$A = 30\text{ft}^2$	Perform given operations
The area is 30 square feet	Write a sentence.

◆**Example 4** Find the base of a triangle whose sides are 8 m and 13 m and whose height measured from the base is 5 m. The area of the triangle is $45m^2$.

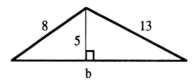

$A = \frac{1}{2}bh$	Formula for area
$90 = \frac{1}{2}(b)(5)$	Substitute
b = 18	Solve
The base is 18 meters.	Write a sentence

Name:_____

8.1 Problem Set

Find the area and perimeter of each rectangle.

1) l = 12 ft w = 8 ft	2) l = 26 cm w = 14 cm	3) l = 38 m w = 24 m

Find the perimeter of each of the following triangles.

4) Sides are 7 ft, 12 ft and 15 ft.	5) Sides are 12 cm, 17 cm, and the third side is 23 cm.	6) The base is 32 m, two equal sides are 23 m each and the height is 12 m.

Find the area of each of the following triangles.

7) b = 42 ft h = 17 ft	8) b = 53 cm h = 24 cm	9) b = 25.4 yds h = 13.7 yds

For each of the following problems, draw and label the figure, write the formula, substitute the given information, solve the equation and write a sentence to answer the question.

10) The length of a rectangle is 15.6 feet. If its perimeter is 58.2 feet, what is its width?	11) The sides of a triangle are 18.4 cm, 22.6 cm and 31.4 cm. The height from the longest side is 12.3 cm. Find the area of the triangle.
12) Ferd Burfil had a front yard that was rectangular in shape. The length was five feet more than twice its width and had a perimeter of 118 feet. He hired Olga to put in new sod lawn. If Olga charged $1.25 per square foot of lawn, how much did she get paid?	13) The Mathematics Study Desk Company had a sale on a desk, charging only six cents per square inch of desk top. If the top was a rectangle 24.3 inches by 66.5 inches, what was the sale price?

8.2 Using Formulas from Geometry: Circles

In this section we will use formulas to solve problems involving circles.

What is a circle?

A **circle** is a closed plane figure made up of a collection of points, everyone of which is equidistant from an inside point called the **center**. The distance from the center to the circle is called the **radius**. A line segment joining any two points of a circle is called a **chord**. A chord that passes through the center is called a **diameter**. Since the diameter consists of two radii, we can say that the diameter is twice the radius, and we write symbolically, d = 2r. In the figure below, C is the center. AC is a radius. AB is a diameter. AD is a chord.

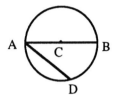

Formulas for a Circle

The distance around a circle is called its **circumference**. The ratio of the circumference of any circle to its diameter is a constant number called **pi** and represented by the Greek letter, π. (For problems in this chapter, you will be using the π key on your calculator or approximating π to be 3.14) This ratio is represented symbolically as :

$$\pi = \frac{C}{d}$$ where C represents the circumference and d represents the diameter.

From this ratio we obtain the formula for circumference,

$$C = \pi d \quad \text{or} \quad C = 2\pi r.$$

The area of a circle is the measurement of the surface bounded by the circle and has the formula,

$$A = \pi r^2$$

◆**Example 1**   Find the area and circumference of a circle having a radius of 8.3 feet.

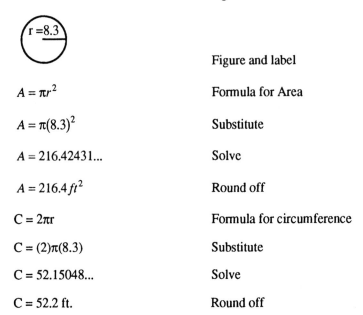

	Figure and label
$A = \pi r^2$	Formula for Area
$A = \pi (8.3)^2$	Substitute
$A = 216.42431...$	Solve
$A = 216.4\, ft^2$	Round off
$C = 2\pi r$	Formula for circumference
$C = (2)\pi(8.3)$	Substitute
$C = 52.15048...$	Solve
$C = 52.2$ ft.	Round off

The area is 216.4 square feet and the circumference is 52.2 feet.

◆**Example 2**   Ferd Burfil wanted to build a circular horse riding ring that would require the same amount of fencing as his current rectangular corral that has a length of 18.4 meters and a width of 12.7 meters. What size radius should he use?.

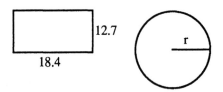

To find the amount of fencing, we must first find the perimeter of the existing corral.

$P = 2l + 2w$

$P = 2(18.4\text{ m}) + 2(12.7\text{ m})$

$P = 36.8\text{ m} + 25.4\text{ m}$

$P = 62.2\text{ m}$

Now set the perimeter of rectangle = the circumference of the circle.

$P = C$

$62.2\text{ m} = 2\pi r$

$r = 9.8994..$

$r = 9.9\text{ m}$

Ferd must build the corral with a radius of approximately 9.9 meters.

Name: _____

Find area and circumference of each of the following circles. Use $\pi = 3.14$ if your calculator does not have a π key.

1) r = 5 inches	2) d = 32 meters	3) r = 4.2 cm

Solve each of the following:

4) Find the area of a circle whose circumference is 57.2 inches.	5) Find the area of a circle whose circumference is 45.9 m.	6) Find the area of a rectangle whose length is 14.6 m and whose perimeter is 46.6 m.

7) Find the base of a triangle whose height is 24.9 feet and whose area is 450.8 sq. feet.	8) Find the area of a circle whose circumference is 28.9 cm.	9) Find the perimeter of a rectangle whose area is 107.1 sq. meters and whose width is 9.0 meters.

Solve each of the following word problems in the usual manner.

10) Ivan had a circular flower bed with a radius of 18.2 feet. If he paid Veronica $1.50 per foot to construct a concrete curb around the edge of the flower bed, how much would Ivan pay Veronica?

11) Olaf designed a silver coin to commemorate the 30th anniversary of the founding of his college. The coin had a diameter of 2.6 cm. How much surface area did the coin have?

12) The Lone Ranger tied his trusty horse Silver to a stake with a rope that was 18 feet long so that the horse could graze in a circular area of Liza's lovely lush lawn while he consumed 3 falafels brimming over with garden fresh tomatoes. How much grazing area did Silver have?

13) The student government sponsored a photo contest to highlight the achievements of the college over the past 30 years. They appropriated prize money totaling $500, with $200 going to the first place winner. Pedro won the contest with a rectangular picture 36 inches long and 24 inches wide. If Pedro spent $28.75 for supplies and paid Fernando 42 cents per inch to frame it, how much profit did Pedro make?

8.3 Using Formulas from Geometry: Pythagorean Theorem

In this section we will use what is probably the most famous formula in mathematics, the Pythagorean theorem. While the Greek mathematician, Pythagoras (580 BC) is generally given credit for discovering the relationship that bears his name, there is strong evidence that Chinese mathematicians were using the concept several hundred years before Pythagoras.

What Is the Pythagorean Theorem?

Before we state the theorem, we need to define a right triangle. A right triangle is a triangle having one right angle (90°). The side opposite the right angle is called the **hypotenuse**. The other two sides are called **legs** of the triangle. Angles are labeled using upper case letters; sides are labeled using lower case letters and have the same letter as the angle opposite the side. For example the side opposite A will be side "a." The side opposite B will be side "b." The side opposite C will be side "c". The square at angle C indicates it is a right angle.

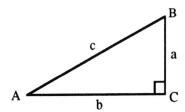

The Pythagorean Theorem states that the square of the hypotenuse of a right triangle equals the sum of the squares of the two legs. If the legs are of length a and b and the hypotenuse is of length c, then we write,

$$a^2 + b^2 = c^2$$

The Principal Square Root Property

In order to solve problems using the Pythagorean Theorem we must understand the concept of square root. We have used the concept of square root in several different sections of this book when we simplified expressions like $\sqrt{25} = 5$ and $\sqrt{49} = 7$. We will now be using the square root concept in equations. We will call this the Principal Square Root Property.

> If each of r and s is a positive number and $r^2 = s$, then $r = \sqrt{s}$ (r is called the principal square root of r or the positive square root of r.)

For example, if $c^2 = 81$, then $c = 9$

If $c^2 = 36$, then $c = 6$

If $c^2 = 14$, then $c = \sqrt{14}$ or $c \approx 3.7$

The symbol ≈ means approximately equal to. Whenever we use the calculator to find the square root of a number that is **not** a perfect square, we must round off the square root to be an approximate answer.

◆ **Example 1** Find the hypotenuse of a right triangle if the legs are 10 yards and 24 yards.

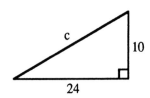

$c^2 = a^2 + b^2$ The Pythagorean Theorem

$c^2 = (10)^2 + (26)^2$

$c^2 = 100 + 576$

$c^2 = 676$

$c = \sqrt{676}$

$c = 26$

The hypotenuse is 26 yards.

◆ **Example 2** Find the hypotenuse of a right triangle if the legs are 12 meters and 17 meters.

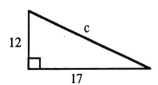

$c^2 = a^2 + b^2$

$c^2 = (12)^2 + (17)^2$

$c^2 = 144 + 289$

$c^2 = 433$

$c = \sqrt{433}$

$c \approx 20.80865$

The hypotenuse is 20.8 meters.

◆**Example 3** Find the length of a leg of a right triangle if the other leg is 12 feet and the hypotenuse is 20 feet.

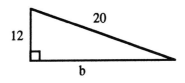

$c^2 = a^2 + b^2$

$(20)^2 = (12)^2 + b^2$

$400 = 144 + b^2$

$256 = b^2$

$\sqrt{256} = b$

$16 = b$

$b = 16$

The other leg is 16 feet.

Name: _____ 8.3 Problem Set 295

Use a calculator to find the indicated root. If the root is not exact, round-off to nearest tenth.

1) $\sqrt{361}$	2) $\sqrt{3136}$	3) $\sqrt{3906.25}$
4) $\sqrt{4673}$	5) $\sqrt{96317}$	6) $\sqrt{76318}$
7) $\sqrt{96.429}$	8) $\sqrt{964.29}$	9) $\sqrt{9642.9}$

For each of the following right triangles, **draw** and **label** the figure, write the **formula**, **substitute** given numbers into the formula and **solve** the equation. On each triangle the hypotenuse is c and the legs are a and b.

10) If a = 5 ft and b = 12 ft find c.	11) If a = 4.5 in and b = 6.0 in find c.	12) If a = 11.2 cm and b = 8.4 cm find c.

8.3 Problem Set

13) If $a = 12.5$ m and $c = 32.5$ m find b.	14) If $b = 16.4$ in and $c = 23.7$ in find a.	15) If $a = 18.4$ ft and $b = 32.9$ ft find c.

Solve the following problems.

16) The base of a ladder is 16.0 feet from the edge of a building. The ladder reaches 38.4 feet up on the building. How tall is the ladder?	17) Given rectangle ABCD with AB = 26.3 inches and BC = 18.6 inches, find the length of AC, which is called the *diagonal* of the rectangle.

18) Given rectangle ABCD with DC = 49.2 cm and the diagonal AC = 63.4 cm, find the length of the side AD.

19) Laurel and Hardy started walking from the same spot at the same time. Laurel walked 5.6 miles due west. Hardy walked 3.7 miles due south. How far are they apart?

20) Romeo has a 24.5 foot ladder. How far from the base of Juliet's house must he place the ladder so that it will exactly reach the base of her window, which is 19.3 feet from the ground?

21) Michael J. Fox and Steve Martin started from the same spot at 8 AM. Mike traveled due east on his skate board at 8 MPH. Steve went due north on his bicycle at 11.6 MPH. How far apart were they at 10 AM?

22) The length of a rectangle is four feet less than three times its width and has a perimeter of 104 feet. Find the area of the rectangle.

23) Architects are slow to learn that students tend to take short cuts. Ivan admired the beautiful rectangular lawn by the math classroom and often strolled on the architect designed walkway surrounding the lawn. One beautiful day he spent too much time "smelling the roses" and chose to take the diagonal route across the lawn rather than the usual walk around the outside. If the lawn measured 120 yards by 80 yards, how much distance did Ivan save by taking the diagonal short cut?

24) In this space, make up and solve an interesting problem using the Pythagorean Theorem.

8.4 Using Geometric Formulas on Irregular Figures

In this section we will turn our organizational skills to problems involving irregular figures. Most irregular figures can be transformed into a combination of regular figures by drawing certain line segments into the irregular figures. The figure below on the left becomes two rectangles by drawing one line segment.

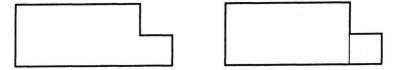

The figure below becomes a rectangle and a triangle by drawing one line segment.

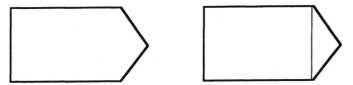

The figure below becomes a rectangle and a half circle by drawing one line segment.

Organization for Irregular Figures

As usual, we need an organized approach to solving problems. Use the following steps to solve problems with irregular figures.

1. **Draw** one or more line segments on the given figure to form two or more regular figures.
2. **Label** the sides of the figure.
3. Write a **plan** to find the total area using areas of the component parts.
4. Decide which **formulas** will be used and write the plan using these formulas.
5. **Substitute** numbers from the figure into the formulas.
6. **Simplify** the expression using the proper order of operations.
7. Write a **sentence**.

◆**Example 1** Find the area and perimeter of the figure below. Each measurement is in feet.

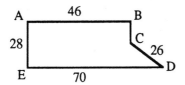

By extending the line segment BC until it intersects the line ED at F, we form a rectangle and a right triangle.

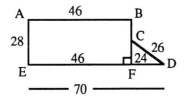

In the above figure, if AB = 46, then EF = 46. Also FD = 70 – 46, so FD is 24. We now need to find the length of CF. Because CF is the leg of a right triangle we will use the Pythagorean Theorem.

$$a^2 + b^2 = c^2$$

$$(CF)^2 + 24^2 = 26^2$$

$$(CF)^2 + 576 = 676$$

$$(CF)^2 = 100$$

$$CF = \sqrt{100}$$

$$CF = 10$$

The segment BC can now be found. BC = 28 – 10. So BC = 18

Plan: Total area is the area of the rectangle added to the area of the triangle.

A = lw + $\frac{1}{2}$bh		Formula
A = (46 ft)(28 ft) + $\frac{1}{2}$ (24 ft)(10 ft)		Substitute
A = 1288 ft^2 + 120 ft^2		Simplify
A = 1408 ft^2		Solve

Plan for the second part of the problem: Perimeter is the sum of all the sides.

P = AB + BC + CD + DE + EF	Formula
P = 46 ft + 18 ft + 26 ft + 70 ft + 28 ft	Substitute
P = 188 ft.	Solve

The figure has an area of 1408 square feet and a perimeter of 188 feet.

◆**Example 2** Find the area and perimeter of the figure below.

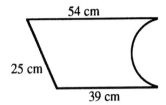

By drawing the figure with two new line segments as shown below, we create three regular figures, namely a triangle, a rectangle and a half circle. In order to label the lengths of the sides we must use algebra. Let x represent one unknown leg of the right triangle formed and let d represent the other leg and the diameter of the half circle. We know the total length of the top is 54 cm, so we solve the algebraic equations below:

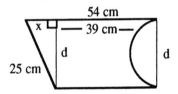

x + 39 cm. = 54 cm.

$\boxed{x = 15 \text{ cm.}}$

and applying the Pythagorean Theorem we can find $d = \sqrt{25^2 - 15^2}$

$\boxed{d = 20 \text{ cm.}}$

Plan: The area of the irregular figure is area of triangle + area of rectangle − area of $\frac{1}{2}$ circle.

$A = \frac{1}{2} bh + lw - \frac{1}{2} \pi r^2$ Formulas

$A = \frac{1}{2} (15 \text{ cm})(20 \text{ cm}) + (39 \text{ cm})(20 \text{ cm}) - \frac{1}{2} (\pi)(10 \text{ cm})$ Substitute

$A = 150 \text{ cm}^2 + 780 \text{ cm.}^2 - 157 \text{ cm.}^2$ Simplify

$\boxed{A = 773 \text{ cm}^2}$

To find the perimeter means to find the distance all the way around the figure. Since one of the sides is a semicircle, we must find half of the circumference before doing the complete perimeter of the figure.

$$\frac{1}{2}C = \frac{1}{2}\pi d \qquad \text{Formula for } \frac{1}{2} \text{ of the circumference}$$

$$\frac{1}{2}C = \frac{1}{2}(\pi)(20 \text{ cm}) \qquad \text{Substitute}$$

$$\frac{1}{2}C = 31 \text{ cm} \qquad \text{(Rounded to nearest whole number)}$$

Plan: Total Perimeter is the sum of all the sides

$$P = a + b + c + \frac{1}{2}C \qquad \text{Formulas}$$

$$P = 54 \text{ cm.} + 25 \text{ cm.} + 39 \text{ cm.} + 31 \text{ cm} \qquad \text{Substitute}$$

$$P = 149 \text{ cm.} \qquad \text{Simplify}$$

The area of the figure is 773 square centimeters and its perimeter is 149 centimeters.

Name:_____   8.4 Problem Set   303

Solve the following problems.

1) Find area and perimeter of the figure below:

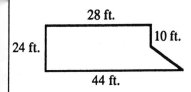

2) Find the area and perimeter of the figure below:

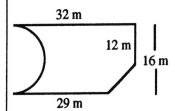

Area Plan:

Formulas:

Substitute:

Solve:

Area Plan:

Formulas:

Substitute:

Solve:

Perimeter Plan:

Formulas:

Substitute:

Solve:

Perimeter Plan:

Formulas:

Substitute:

Solve:

3) Find the area of the figure below:

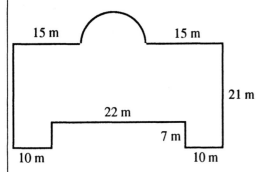

4) Find the area and perimeter of the figure below:

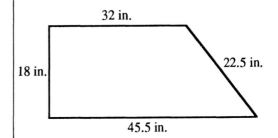

5) Find the area of the shaded part of the figure. Assume that the circles are **tangent** to each other and to the sides of the rectangle. (Two figures are **tangent** if they touch in exactly one point.)

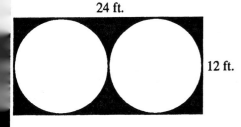

24 ft.

12 ft.

6) A circle with a diameter of 10 inches contains a rectangle whose vertices (corners) are on the circle. The width of the rectangle is 6 inches. Find the area outside the rectangle but inside the circle.

7) As part of her dance routine, Isadora placed the foot of a ladder 9.4 feet from the bottom of a wall. If the ladder reached 26.8 feet high on the wall, how long was the ladder?

8) The diagonal of a square is 32.7 cm long. How long is the side of the square?

9) Make up and solve a problem involving an irregular figure.

8.5 Dealing with Dimensions

The problems in this chapter have dealt with measurement numbers. That is, we have been given information about the size of geometric figures using both a number and a unit of measure. As we measure the length of a rectangle, we might say that the length is 9 feet or 9 inches or 9 meters. The measurements used in geometry involve two parts, the numerical part and the unit of measure part. In this section, we will learn how to change from one unit of measure to another unit of measure.

Conversion of Units of Measure

When it is necessary to convert from one unit of measure to another, we use a familiar pattern of mathematics, namely multiplying by the identity element "1" As we saw in the study of fractions, when a number is multiplied by "1" the value stays the same. Canceling like terms is yet another use of the identity element "1." Whenever a number is represented as a fraction whose numerator has the same value as the denominator, we say the number is "1", provided that the denominator is not zero.

Organization for Conversion of Units of Measure

The following organization will be used in converting units of measure.

1. Write the equivalents to be used in the problem. Use appendix 1 if needed.
2. Express the equivalents as fractions equal to one.
3. Use these equivalents to re-write the original unit of measure into the desired unit measure.
4. Use the concept of "canceling" to express the measurement into the desired measurement.

◆ **Example 1** Convert 3 gallons to pints.

$2 \text{ pints} = 1 \text{ quart}$

$4 \text{ quarts} = 1 \text{ gallon}$ Equivalents to be used

$3 \text{ gallons} = \dfrac{3 \text{ gallons}}{1} \cdot \dfrac{4 \text{ quarts}}{1 \text{ gallon}} \cdot \dfrac{2 \text{ pints}}{1 \text{ quart}}$ Express as fractions = to one

$= \dfrac{3 \cancel{\text{ gallons}}}{1} \cdot \dfrac{4 \cancel{\text{ quarts}}}{1 \cancel{\text{ gallon}}} \cdot \dfrac{2 \text{ pints}}{1 \cancel{\text{ quart}}}$ Cancel

$= 24 \text{ pints}$ Simplify

| 3 gallons is the same as 24 pints. | Sentence

Helpful Hint 17: *In using the equivalents, build the fraction so that the unwanted units cancel out. In the above example, the gallons in the numerator cancel with the gallons in the denominator; the quarts in the numerator cancel with the quarts in the denominator. When all the cancellations are done, the only remaining unit is pints which was the desired unit.*

Chapter 8: Geometry

♦**Example 2** Convert 2642 ounces into quarts and round answer to nearest tenth of a quart.

From appendix 1 we find the following equivalents:

8 fluid ounces = 1 cup, so the fraction, $\dfrac{1 \text{ cup}}{8 \text{ oz.}} = 1$

2 cups = 1 pint, so the fraction, $\dfrac{1 \text{ pint}}{2 \text{ cups}} = 1$

2 pints = 1 quart, so the fraction, $\dfrac{1 \text{ quart}}{2 \text{ pints}} = 1$

Now we will use these forms of the number 1 to rewrite the original problem.

$$2642 \text{ oz.} = \frac{2642 \text{ oz}}{1} \cdot \frac{1 \text{ cup}}{8 \text{ oz}} \cdot \frac{1 \text{ pt}}{2 \text{ cup}} \cdot \frac{1 \text{ qt}}{2 \text{ pt}}$$

$$= \frac{2642 \text{ qt}}{32}$$

$$= 82.5625 \text{ quarts (which rounds off to the nearest tenth to be 82.6)}$$

2642 ounces is approximately the same as 82.6 quarts.

The Meaning of Rate

The concept of rate is often used. You may drive your car 60 miles per hour. You may be able to type 50 words per minute. You may earn $8 per hour. Each of these is an example of rate. A rate is the ratio of some measurable event compared to some unit of time. The rate of travel of 60 MPH is the distance traveled in miles divided by the time required in hours.

$$r = \frac{d}{t}$$

♦**Example 3** A runner won a 100 yard dash by running a time of 9.8 seconds. What was her time in miles per hour, rounded to the nearest tenth of a mile per hour.

From Appendix 1 we will use the following equivalents for this problem.

1 mile = 1760 yards

1 minute = 60 seconds

1 hour = 60 minutes

$$100 \text{ yards in 9.8 seconds} = \frac{100 \text{ yds}}{9.8 \text{ sec}} \cdot \frac{1 \text{ mile}}{1760 \text{ yds}} \cdot \frac{60 \text{ sec}}{1 \text{ min}} \cdot \frac{60 \text{ min}}{1 \text{ hr}}$$

$$= \frac{360000 \text{ miles}}{17248 \text{ hours}}$$

$$= 20.871985 \text{ miles per hour}$$

Running 100 yards in 9.8 seconds is the same as running 20.9 miles per hour.

Use dimensional analysis to make the following conversions.

1) Convert 28 yards to inches.	2) Convert 3 weeks to hours.
3) Convert 5 square yards to square feet.	4) Convert 62 inches to yards. (Round to nearest tenth of a yard.)
5) Convert 7.2 acres to square feet.	6) Convert 510 ounces to pounds. (Round to nearest tenth of a pound.)
7) Convert $2\frac{3}{4}$ miles to yards.	8) Convert 2 weeks and 3 days to hours.
9) Convert 15,680 ft² to acres. (Round to nearest hundredth of an acre.)	10) Convert 1365 yd² into acres (Round to the nearest hundredth of an acre.)
11) Convert 185 pounds of fresh water into gallons.	12) Convert 18,750 ounces into tons (Round to the nearest thousandth of a ton.)

8.5 Problem Set

13) The Lone Ranger's faithful horse, Silver can run 1000 yards in a blistering 2 minutes. What is Silver's speed (to the nearest tenth) in miles per hour?

14) Olga has a pump that can fill a tank at the rate of 18 quarts per minute. What is the rate of her pump in gallons per hour?

15) The Batmobile has been clocked at 112 miles per hour on a legal test track. How fast (to the nearest tenth) is that in feet per second?

16) Sean Patrick Murphy can go 60 yards on his skate board in a mere 7.8 seconds. How fast is that in miles per hour, rounded to the nearest tenth.

17) Garth Brooks can plow 150 square yards per minute. What is his rate in acres per hour, rounded to the nearest hundredth.

18) If the Batmobile gets 18 MPG, what is that in feet per quart?

8.6 Chapter Review

In this chapter you have learned to:

1. Use geometric formulas to solve problems involving rectangles, triangles and circles.
2. Apply the Pythagorean Theorem to solve right triangle problems.
3. Use a combination of formulas to solve irregular figures.
4. Use dimensional analysis to convert units of measure.

Geometric Formulas Used   [8.1, 8.2, 8.3]

$A = lw$	Area of a rectangle
$P = 2l + 2w$	Perimeter of a rectangle
$A = \frac{1}{2} bh$	Area of a triangle
$P = a + b + c$	Perimeter of a triangle
$A = \pi r^2$	Area of a circle
$C = \pi d$ or $C = 2\pi r$	Circumference of a circle
$a^2 + b^2 = c^2$	Pythagorean Theorem

Organization Used in Solving a Geometry Problem   [8.1]

1. **Draw** and **label** the figure to represent the problem.
2. Write the **formula** to be used.
3. **Substitute** given numbers in the formula. (Remember empty parenthesis)
4. **Perform** the given operations or **solve** the resulting equation.
5. Write a **sentence** to answer the question.

◆**Example 1**   Find the perimeter of a rectangle whose area is 242.5 ft² and whose length is 19.4 ft.

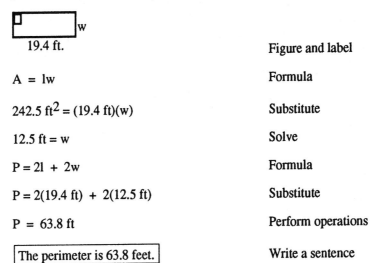

19.4 ft.	Figure and label
$A = lw$	Formula
$242.5 \text{ ft}^2 = (19.4 \text{ ft})(w)$	Substitute
$12.5 \text{ ft} = w$	Solve
$P = 2l + 2w$	Formula
$P = 2(19.4 \text{ ft}) + 2(12.5 \text{ ft})$	Substitute
$P = 63.8 \text{ ft}$	Perform operations
The perimeter is 63.8 feet.	Write a sentence

Using the Pythagorean Theorem [8.3]

The Pythagorean Theorem is used in finding a side of a right triangle. The side opposite the right angle is called the hypotenuse while the other two sides are called legs.

◆**Example 2**   If the hypotenuse of a right triangle is 32.5 m and one leg is 26.0 m, find the other leg.

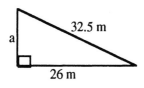

$a^2 + b^2 = c^2$	The Pythagorean Theorem
$a^2 + (26)^2 = (32.5)^2$	Substitute
$a^2 + 676 = 1056.25$	Perform indicated operations
$a^2 = 380.25$	Simplify
$a = \sqrt{380.25}$	
$a = 19.5$	
The other leg is 19.5 meters.	Sentence

Organization for Problems Having Irregular Figures   [8.4]

As usual, we need an organized approach to solving problems. Use the following steps to solve problems with irregular figures.

1. Draw one or more line segments on the given figure to form two or more regular figures.
2. Label the sides of the figure.
3. Write a plan to find the total area using areas of the component parts.
4. Decide which formulas will be used and write the plan using these formulas.
5. Substitute numbers from the figure into the formulas.
6. Simplify the expression using the proper order of operations.
7. Write a sentence.

◆**Example 3** Find the area of the figure below:

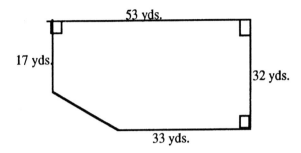

First draw dotted lines to make regular figures. Extend left side down and let x represent its length. Extend bottom to the left and let a represent its length. A right triangle is formed by these two extensions. This also forms a rectangle.

$17 + x = 32$ Left side $(17 + x)$ = the right side (32)

$\boxed{x = 15}$

$a + 33 = 53$ Bottom side $(a + 33)$ = top side (53)

$\boxed{a = 20}$

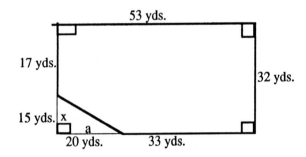

Total Area = Area of Rectangle − Area of right triangle.   Plan

$TA = lw - \frac{1}{2}bh$ Formulas

$TA = (53 \text{ yds})(32 \text{ yds}) - \frac{1}{2}(15)(20)$ Substitute

$TA = 1696 \text{ yds}^2 - 150 \text{ yds}^2$ Operate

$TA = 1546 \text{ yds}^2$ Solve

$\boxed{\text{The total area is 1546 square yards.}}$ Sentence

Organization for Conversion of Units of Measure

The following organization will be used to convert units of measure.

1. Write the equivalents to be used in the problem. Use appendix 1 if needed.
2. Express the equivalents as fractions equal to one.
3. Use these equivalents to re-write the original unit of measure into the desired unit of measure.
4. Use the concept of "canceling" to express the original unit of measure into the desired unit of measure.

◆**Example 4** A pump can fill a tank at the rate of 38 quarts per minute. Find the rate of filling the tank in gallons per hour.

$$1 \text{ gallon} = 4 \text{ quarts}$$

$$1 \text{ hour} = 60 \text{ minutes} \qquad \text{Equivalents to be used.}$$

$$\frac{38 \text{ qts}}{1 \text{ min}} = \frac{38 \text{ qts}}{1 \text{ min}} \cdot \frac{60 \text{ min}}{1 \text{ hr}} \cdot \frac{1 \text{ gallon}}{4 \text{ qts}} \qquad \text{Express equivalents as fractions} = 1$$

$$= \frac{38 \text{ qts}}{1 \text{ min}} \cdot \frac{60 \text{ min}}{1 \text{ hr}} \cdot \frac{1 \text{ gallon}}{4 \text{ qts}} \qquad \text{Cancel}$$

$$\frac{38 \text{ qts}}{1 \text{ min}} = \frac{570 \text{ gallons}}{1 \text{ hour}} \qquad \text{Simplify}$$

| The pump can fill the tank at the rate of 570 gallons per hour. | Sentence

Name: _____   8.6 Chapter Review Test

Solve for the indicated value.

1) Find the area of a triangle whose base is 12.8 yards and whose height is 6.5 yards.	2) Find the area of a circle whose diameter is 8.4 feet.	3) Find the width of rectangle that has a perimeter of 49.2 cm. and a length of 15.2 cm.
4) The hypotenuse of a right triangle is 31.2 meters and one leg is 12.0 meters. Find the other leg.	5) A right triangle has an area of 37.5 ft.^{2} and a base of 7.5 feet. Find its height.	6) Find the circumference of a circle with a radius of 9.3 feet.

Solve the following problems in the usual manner.

7) Find the perimeter of a right triangle whose legs are 15.2 feet and 21.6 feet.	8) Find the area of a right triangle whose base is 9.3 feet and whose hypotenuse is 15.5 feet.

9) The length of a rectangle is one yard more than three times its width. If the perimeter is 122 yards, what is its length?

10) If you can run a 440 yard dash in 58.6 seconds, how fast are you running in miles per hour? (Give answer to the nearest tenth of a mile per hour.)

11) A rectangular shaped garage door measures 16.4 feet long and 8.7 feet high. If one quart of paint covers approximately 90 sq. ft., how many quarts of paint would you need to buy to paint the garage door?

12) Ferd made a triangular kite with a base of 18.4 inches and a height of 25.2 inches. Fannie made a rectangular kite with a length of 16.2 inches and a width of 14.2 inches. Which has more area?

13) The figure below is a drawing of a small back yard with dimensions given in feet. Find the cost of planting a sod lawn if the cost is 84 cents per square foot.

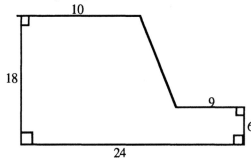

Name: _____

Evaluate using $a = -2$ $b = 6$ $n = \dfrac{2}{3}$ $s = \dfrac{3}{4}$ $x = -8$

1) $nab - x$	2) $ns + ab$	3) $sb^2 - a^2 s$
4) $nxb + \sqrt{b+a}$	5) $\sqrt{-abs} - sx$	6) $(b-a)^2 + bnx$

Solve each of the following equations.

7) $7x - 2 + 4x = 5x + 2$	8) $3 + 7(x - 5) = 3x + 8$	9) $8(2x + 4) - 5(2x - 8) = 0$

9.1 Final Review Problem Set

Name:_____

Evaluate each function for the given value.

10) If $f(x) = 5x - 2$ Find: $f(6) =$ $f(-4) =$ $f\left(\dfrac{4}{5}\right) =$ $f\left(\dfrac{3}{4}\right) =$ $f(0) =$	11) If $g(x) = 6x + 5$ Find $g(2) =$ $g(-3) =$ $g\left(\dfrac{3}{4}\right) =$ $g\left(-\dfrac{2}{3}\right) =$ $g(0) =$
12) If $h(x) = 4x^2 - 3x + 2$ Find: $h(3) =$ $h(-5) =$ $h(6) =$ $h(0) =$	13) If $R(x) = 6x + \sqrt{3x - 1}$ Find $R\left(\dfrac{1}{3}\right) =$ $R\left(\dfrac{2}{3}\right) =$ $R\left(\dfrac{5}{3}\right) =$ $R\left(\dfrac{10}{3}\right) =$

14) The larger of two numbers is seven more than twice the smaller. Six times the smaller increased by three times the larger is 129. Find the sum of the two numbers.

15) Tarzan is six years older than three times the age of Fannie Farkel. Twice Tarzan's age increased by three times Fannie's age is 138. Find the age of Boris who is five years older than Tarzan.

16) Two-thirds of a number decreased by seven is the same as one-half the number increased by twelve. What is the number?

17) The length of a rectangle is three more than twice its width. If the perimeter is 72 feet, find its area.

18) Add $2\frac{5}{8}$ to the product of $1\frac{1}{6}$ and $3\frac{3}{4}$.

19) Olaf was paid $5 per day plus a commission of five-eighths of all his sales. Write a function to describe his pay. Use the sample sales figures with x representing his sales and graph the function.

$$P(x) =$$

P(16) =   P(40) =   P(56) =   p(88) =

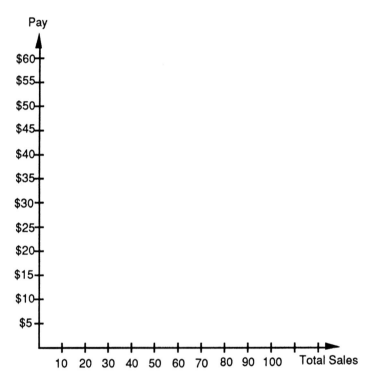

Use the above graph to answer the following questions: (DO NOT CALCULATE)

20) Find P(8) =

21) Find P(72) =

22) If P(x) = 25 find the value of x.

23) If P(x) = 55 find the total amount of Olaf's sales.

Name: _____   9.2 Final Review Problem Set

Find the missing values.

1) 12% of _____ is 7.8

2) _____ % of 36 is 1800

3) 46% of 365 is _____

4) 3.2% of _____ is 7.84

5) 125% of 76 is _____

6) 18 is _____ % of 80

Perform the indicated operation.

7) $7\frac{3}{8} + 2\frac{5}{6} - 4\frac{3}{4}$

8) $\left(1\frac{1}{6}\right)\left(\frac{3}{10}\right) + \frac{5}{6}$

9) $\frac{2}{3} + 1\frac{1}{3}\left(7 - 2\frac{1}{2}\right)$

Perform the indicated operation.

10) $7\frac{5}{21} + 3\frac{3}{14} - 2\frac{1}{2}$

11) $8\frac{7}{24} - 5\frac{3}{16} + 2\frac{5}{6}$

12) Tarzan rode 13 miles on his elephant, rode 18 miles on his bike and walked 9 miles to visit his 26 year old friend, Boris. What percent of his total distance was on the bike?

13) Fannie Farkel bought a TV set that had a regular price of $520. She was given a special "math student" discount of 25% off the regular price but still had to pay the state sales tax of 8.5%. What was her total purchase price?

14) Boris was paid $6 per hour plus 8% commission on sales that he made. What was his total earnings the day he worked from 3 PM to 7 PM and sold a total of $280?

15) Tarzan made one of his famous jungle salads using $5\frac{2}{3}$ pounds of lettuce, $2\frac{3}{4}$ pounds of tomatoes, $1\frac{1}{3}$ pound of onions and a half pound of garlic. How many pounds of nuts should he add so that the salad weighs exactly 12 pounds?

Name: _____

16) Subtract $1\frac{4}{5}$ from the product of $6\frac{1}{2}$ and $\frac{2}{5}$.

17) Multiply $5\frac{1}{3}$ by the sum of $\frac{7}{8}$ and $\frac{3}{4}$.

18) Ivan took a vacation and spent twice as much for his hotel as he did for his airfare. His entertainment and food were $75 more than the cost of the airfare. If the total cost of the trip was $1235, how much did he spend on the hotel?

19) The sum of two numbers is 42. If six times the smaller number is decreased by four times the larger, the result would be 22. Find the two numbers.

20) The width of a rectangle is 12.8 yards. Find the length if the perimeter is 72.4 yards.

21) Find two numbers whose sum is six. such that six times the smaller increased by three times the larger is exactly 25.

Solve or simplify

22) $6 + 3(x - 5) = 45$

23) $48 = 8 - 4(x - 5)$

24) $8(2 - x) = 2(2x - 10)$

25) $\frac{2}{3}(x + 12) = 24$

26) $\frac{3}{8}x + 6 + \frac{1}{2}x = 34$

27) $6 + 5(2x - 3) = 2(1 - 2x$

28) $2\frac{1}{2} + 3\frac{3}{7} \div \frac{9}{14}$

29) $\left(1\frac{3}{5}\right) \cdot \left(\frac{5}{8}\right)^2 + 1\frac{3}{4}$

30) $7\frac{5}{8} + 2\frac{7}{12} - 4\frac{11}{15}$

Solve or simplify.

1) $4 + 7(x - 4) = 6 - 4(6 - x)$	2) $3x - 7 = 4(2x - 3) - 20$	3) $6 - 3(x - 4) - 5(x + 2) = 0$
4) 125% of _____ is 70	5) 240 is _____ % of 30	6) 3.5% of 7400 is _____
7) $5\frac{2}{3} + \left(6\frac{1}{4}\right) \cdot \left(1\frac{1}{15}\right)$	8) $\left(1\frac{7}{8}\right) \cdot \left(\frac{2}{3}\right)^2 + 8\frac{1}{4} \div 4\frac{1}{2}$	9) $\dfrac{\sqrt{16+9} + (2)(5)}{(4)(7) - 5^2}$

Solve or simplify.

10) $5\frac{9}{14} + 2\frac{10}{21} - 4\frac{5}{6}$ 14 = 21 = 6 = LCD = Change to LCD	11) $6\frac{5}{8} + 3\frac{4}{15} - 5\frac{7}{12}$ 8 = 15 = 12 = LCD = Change to LCD
12) Find two numbers whose sum is 25 such that seven times the larger increased by four times the smaller is 151.	13) If the length of a rectangle is 12.8 feet and the perimeter is 40.8 feet, what is the width?
14) Find the result when the product of $1\frac{3}{4}$ and $2\frac{2}{3}$ is subtracted from ten.	15) Wilbert purchased a TV set which normally sells for $350. He used his "Good Student" discount of 12% but still had to pay the 8.5% state sales tax. What was his total cost?

16) Subtract the quotient of $3\frac{3}{5}$ and $2\frac{1}{4}$ from $8\frac{4}{5}$.

17) Wanda earned $7 per hour plus 6% commission for sales that she made. How much did she earn the day she worked from 1 PM to 6 PM and sold a total of $312 worth of goods?

18) Boris bought new size $9\frac{1}{2}$ running shoes to help celebrate his arrival at age 21. The shoes had a regular price of $75 but they were on sale at a 25% discount. What was his total cost if the sales tax was 8.5%?

19) Gilda used Tarzan's special salad for the dinner she had to help celebrate Boris's birthday. She used $2\frac{5}{6}$ pounds of tomatoes, $3\frac{1}{2}$ pounds of Romaine lettuce, $1\frac{1}{3}$ pounds of spinach and $\frac{2}{3}$ pounds of fresh garlic from Gilroy. The party started at 5:30 PM and by 8 PM Gilda and her nine guests had eaten the entire salad. On the average, how much did each person eat?

20) Find the area and circumference of a circle whose radius is 4.6 inches.

21) In a recent bike race Gilda averaged 20 MPH. What was her average rate in feet per second?

22) Ferd charges 65 cents per square foot to plant a sod lawn. Fannie charges $3.18 per foot to build a concrete mowing strip. If Tom Hanks hired them to build the mowing strip and plant the lawn on a rectangular plot that was 72 feet long and 48 feet wide, how much did it cost him?

23) Michael Jordan wanted to carpet the room shown in the figure below. The measurements are given in feet but the carpet is sold by the square yard. If the installed price is $46 per square yard, what did the carpet cost him?

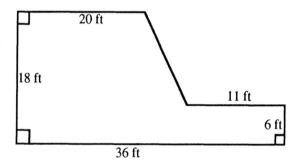

Weights and Measures

Length

12 inches = 1 foot

3 feet = 1 yard

5280 feet = 1 statute mile

1760 yards = 1 statute mile

Time

60 seconds = 1 minute

60 minutes = 1 hour

24 hours = 1 day

7 days = 1 week

360 days = 1 business year

365 days = 1 year

12 months = 1 year

52 weeks = 1 year

Area

144 square inches = 1 square foot

9 square feet = 1 square yard

43,560 square feet = 1 acre

640 acres = 1 square mile

Volume

1,728 cubic inches = 1 cubic foot

27 cubic feet = 1 cubic yard

Dry Capacity

2 pints = 1 quart

8 quarts = 1 peck

4 pecks = 1 bushel

Liquid Capacity

8 fluid ounces = 1 cup

2 cups = 1 pint

2 pints = 1 quart

4 quarts = 1 gallon

Weight

16 ounces = 1 pound

2240 pounds = 1 long ton

2000 pounds = 1 ton

Equivalents of Weight, Volume, and Capacity

231 cubic inches = 1 gallon

7.5 gallons = 1 cubic foot

62.5 pounds = 1 cu. ft fresh water

64 pounds = 1 cu. ft salt water

Useful Formulas

Distance

$d = rt$ — Find distance, given rate and time

$r = \dfrac{d}{t}$ — Find rate, given distance and time

$t = \dfrac{d}{r}$ — Find time, given distance and rate

Temperature

$F = \dfrac{9}{5} C + 32$ — Convert from Celsius to Fahrenheit

$C = \dfrac{5}{9} [F - 32]$ — Convert from Fahrenheit to Celsius

Rectangle

$A = lw$ — Area

$P = 2l + 2w$ — Perimeter

Triangle

$A = \dfrac{1}{2} bh$ — Area

$P = a + b + c$ — Perimeter

$c^2 = a^2 + b^2$ — Sides of a right triangle with c = hypotenuse

Circle

$A = \pi r^2$ — Area

$C = 2\pi r$ — Circumference

$d = 2r$ — Relation of diameter to radius

Rectangular, Solid

$V = lwh$ — Volume

$SA = 2[lw + wh + hl]$ — Surface area

Cylinder

$V = \pi r^2 h$ — Volume

$SA = 2\pi r^2 h + 2\pi rh$ — Surface area

Cone

$V = \dfrac{1}{3} \pi r^2 h$ — Volume

1.1 Answers

1) $8 + (2)(3) - 5$
   $8 + 6 - 5$
   $14 - 5$
   $\boxed{9}$

4) $6 + 35 \div 7 - 2$
   $6 + 5 - 2$
   $11 - 2$
   $\boxed{9}$

7) $6^2 + 3$
   $36 + 3$
   $\boxed{39}$

10) $7(2)^2 - 2(3)^2$
    $7(4) - 2(9)$
    $28 - 18$
    $\boxed{10}$

13) $\frac{21}{7} + \frac{45}{9}$
    $3 + 5$
    $\boxed{8}$

16) $12 + 4^2[3^2 - 2^3]$
    $12 + 16[9 - 8]$
    $12 + 16[1]$
    $12 + 16$
    $\boxed{28}$

19) $ba - xw$
    $(12)(5) - (3)(2)$
    $60 - 6$
    $\boxed{54}$

22) $ax + bw$
    $(5)(3) + (12)(2)$
    $15 + 24$
    $\boxed{39}$

25) $b + a[xb - aw]$
    $(12) + (5)[(3)(12) - (5)(2)]$
    $12 + 5[36 - 10]$
    $12 + 5[26]$
    $12 + 130$
    $\boxed{142}$

28) Fernando bought three pineapples at a fruit stand in Hawaii. If each pineapple cost $2 and he gave the clerk a new $20 bill, how much change should he get?

    $20 - 3(2)$
    $= 20 - 6$
    $= \boxed{14}$

    Fernando got $14 in change.

30) If you had bowling scores of 129, 152, 131 and 144, what would your average score be?

    $\frac{129 + 152 + 131 + 144}{4}$
    $= \frac{556}{4}$
    $= \boxed{139}$

    Your average score would be 139.

2) 13   5) 6   8) 8   11) 7   14) 5   17) 1
20) 21   23) 27   26) 15   29) $15   31) $14

1.2 Answers

1) $-5+8=$ 3
2) $-6+(-2)=$ -8
3) $-16+10=$ -6
4) $-13+18=$ 5
5) $-26+(-2)=$ -28
6) $7-2=$ 5
7) $3-8=$ -5
8) $4-(-3)=$ 7
   $4+3$
9) $-2-(-5)=$
   $-2+5=3$
10) $-6-4=$ -10
11) $(-6)(3)=$ -18
12) $(-4)(-3)=$ 12
13) $(-12)\div(-2)=$ 6
14) $(-15)\div 5=$ -3
15) $18\div(-3)=$ -6

---

16) $(-6)(2)+(-4)(5)$
    $-12 + -20$
    $\boxed{-32}$

19) $\frac{-16}{-2} - \frac{-14}{-2}$
    $8 - 7$
    $\boxed{1}$

22) $\frac{-16}{-2}+(-6)(4)$
    $= 8 + (-24)$
    $= \boxed{-16}$

25) $(-3)^2 + 4(-2)^3$
    $9 + 4(-8)$
    $9 + -32$
    $\boxed{-23}$

28) akxw
    $(3)(-2)(-4)(5)$
    $(-6)(-20)$
    $\boxed{120}$

31) ax + kn
    $(3)(-4)+(-2)(-6)$
    $-12 + 12$
    $\boxed{0}$

34) $xn - k[w-n]$
    $= (-4)(-6)-(-2)[(5)-(-6)]$
    $= 24 + 2[5+6]$
    $= 24 + 2[11]$
    $= 24 + 22$
    $= \boxed{46}$

37) $kx[n-a]+7n$
    $(-2)(-4)\bigl[(-6)-(3)\bigr]+7(-6)$
    $8[-9] + (-42)$
    $-72 + (-42)$
    $\boxed{-114}$

40) Boris bought 16 jars of borscht at $2 per jar. He paid with a $100 travelers check. **How much change did he receive?**

    $\$100 - 16(\$2)$
    $\$100 - \$32$
    $\boxed{\$68}$

    Boris received $68 in change.

42) Waldo walked for a solid five hours, averaging 4 MPH and then rode his bike at a steady 12 MPH for the next three hours. Find the total **distance that Waldo traveled**. (Use the formula: distance = rate times times)

    $5(4) + 3(12)$
    $20 + 36$
    $\boxed{56}$

    Total distance Waldo traveled was 56 miles.

---

17) -32 20) -22 23) -11 26) 15   32) -18 35) -46

38) -6 41) $672   43) $68

ADDITION

1) $(-9) + 3 = -6$
2) $(-7) + (-5) = -12$
3) $8 + (-13) = -5$
4) $-6 + 15 = 9$
5) $9 + (-18) = -9$

SUBTRACTION

6) $6 - 8 = -2$
7) $8 - 6 = 2$
8) $-2 - (-7) = 5$
   $-2 + 7$
9) $-3 - 4 = -7$
10) $-9 - (-5) = -4$
    $-9 + 5$

MULTIPLICATION & DIVISION

11) $(-8)(7) = -56$
12) $(-3)(-8) = 24$
13) $\frac{42}{-7} = -6$
14) $-81 \div (-9) = 9$
15) $(-2)(-3)(-4) = -24$

---

16) $(-5)(-3) + (-4)(-2)$
    $15 + 8$
    $\boxed{23}$

19) $\frac{-32}{4} + (-2)^3$
    $-8 + (-8)$
    $\boxed{-16}$

22) $-6 + 3(-4) - (-5)(2)$
    $-6 + (-12) - (-10)$
    $-18 + 10$
    $\boxed{-8}$

25) $-8 + (-2)^4(3) - \frac{12}{-3}$
    $-8 + 16(3) - (-4)$
    $-8 + 48 + 4$
    $40 + 4$
    $\boxed{44}$

---

28) $ak + nx$
    $(4)(-3) + (-2)(6)$
    $-12 + (-12)$
    $\boxed{-24}$

31) $a^2 - k^2[n-x]$
    $(4)^2 - (-3)^2[(-2)-(6)]$
    $16 - 9[-8]$
    $16 - (-72)$
    $16 + 72$
    $\boxed{88}$

34) $[n+x] \div a + z$
    $[(-2)+(6)] \div (4) + (2)$
    $4 \div 4 + 2$
    $1 + 2$
    $\boxed{3}$

37) $nx \div a + z$
    $(-2)(6) \div (4) + (2)$
    $-12 \div 4 + 2$
    $-3 + 2$
    $\boxed{-1}$

---

40) Myron took a vacation and had the following expenses: four nights lodging at $48 per night, 4 days car rental at $34 per day, airline tickets at $248, food at $108. The remainder of his $882 total expense was spent on entertainment. How much did he spend on entertainment?

    $4(48) + 4(34) + 248 + 108$
    $192 + 136 + 248 + 108$
    $328 + 356$
    $\boxed{\$684}$ TOTAL non entertain

    $\$882 - 684 =$ For entertainment
    $= \$198$

    Myron spent $198 on entertainment.

42) Arnold went to the race track and won $15 on the first race, then lost $8 on each of the next four races. His luck changed on the sixth race when he won $12. What was his total financial result on these six bets?

    $15 + 4(-8) + 12$
    $15 + (-32) + 12$
    $-17 + 12$
    $\boxed{-5}$

    Arnold lost $5 at the races.

---

17) 13   20) 58   23) −57   26) 43   29) 52   32) 35

35) 72   38) 0   41) $94   43) 35

1.4 Answers

ADDITION

1) $-2 + (-7) = -9$
2) $-3 + (-8) = -11$
3) $-15 + (-9) = -24$
4) $-16 + 9 = -7$
5) $-18 + 3 = -15$

SUBTRACTION

6) $9 - 12 = -3$
7) $-6 - 3 = -9$
8) $-9 - (-4) = -5$
   $-9 + 4$
9) $8 - (-3) = 11$
   $8 + 3$
10) $2 - 14 = -12$

MULTIPLICATION & DIVISION

11) $(-6)(8) = -48$
12) $\frac{-4}{-2} = 2$
13) $(-15)(-3) = 45$
14) $\frac{-75}{3} = -25$
15) $(-5)(2)(-3) = 30$

16) $(-8)(-6) + (-7)(2)$
    $48 + -14$
    $\boxed{34}$

19) $\frac{-36}{-6} + (-6)^2 + 3^2$
    $6 + 36 + 9$
    $42 + 9$
    $\boxed{51}$

22) $(-2)(4)^2 + 3(-5)^2$
    $-2(16) + 3(25)$
    $-32 + 75$
    $\boxed{43}$

25) $-8 + 3(-6) - \frac{27}{-3}$
    $-8 + (-18) - 9$
    $-26 - 9$
    $\boxed{-35}$

28) $vn - sa$
    $(-3)(3) - (1)(-4)$
    $-9 - (-4)$
    $-9 + 4$
    $\boxed{-5}$

31) $nxb - v$
    $(3)(5)(-2) - (-3)$
    $-30 + 3$
    $\boxed{-27}$

34) $a^2 + vs^3$
    $(-4)^2 + (-3)(1)^3$
    $16 + (-3)(1)$
    $16 + (-3)$
    $\boxed{13}$

37) $nb[nv + x] + a$
    $(3)(-2)[(3)(-3) + (5)] \div (-4)$
    $-6[-9 + 5] \div (-4)$
    $-6[-4] \div (-4)$
    $24 \div (-4)$
    $\boxed{-6}$

50) Find the result when five more than seven times six is added to the product of six and three.
    $5 + 7(6) + 6(3)$
    $5 + 42 + 18$
    $47 + 18$
    $\boxed{65}$
    The result is 65.

52) Find the result when the square of four is added to the quotient of six and two.
    $4^2 + \frac{6}{2}$
    $16 + 3$
    $\boxed{19}$
    The result is 19.

54) Find the results when the product of nine and six is diminished by three times the sum of four and six.
    $(9)(6) - 3(4+6)$
    $54 - 3(10)$
    $54 - 30$
    $\boxed{24}$
    The result is 24.

17) 57   20) –110   23) 0   26) 38   29) –32   32) –18   35) –6

38) 120   51) –39   53) –75   55) –22

1) $(-4)(6) + (-5)(-4)$

$-24 + 20$

$\boxed{-4}$

4) $\dfrac{-7-3}{-8+3} + \dfrac{-24}{4}$

$\dfrac{-10}{-5} + (-6)$

$2 + (-6)$

$\boxed{-4}$

Evaluate using $a = -5$

7) $ab - cx$

$(-5)(8) - (-2)(-3)$

$-40 - (6)$

$\boxed{-46}$

10) $\dfrac{c+x}{a} + \dfrac{bx}{c}$

$\dfrac{(-2)+(-3)}{(-5)} + \dfrac{(8)(-3)}{(-2)}$

$\dfrac{-5}{-5} + \dfrac{-24}{-2}$

$1 + 12$

$\boxed{13}$

22) Abdul enjoyed a full week in Hawaii. He spent $82 per night on his hotel, $45 per day for food, and an average of $38 per day for miscellaneous expenses. If his airfare cost him $429, and he budgeted $1800 for the trip, **how much did he have remaining** to buy a gift for his hard working mathematics teacher.

$\$82(7) + \$45(7) + \$38(7) + \$429$

$\$574 + \$315 + \$266 + \$429$

$\$889 + \$695$

$\boxed{\$1584 \text{ TOTAL EXP}}$

$\$1800 - \$1584 = \$216$

Abdul had $216 remaining for a Mathematics teacher gift.

24) Sherlock lost $4 per share on 125 shares of IBM stock but gained $6 per share on 180 shares of Ford stock. **What was his net profit** or loss if he had to pay the stock broker a total of $172 in commissions.

$125(-4) + 180(6) - 172$

$-500 + 1080 - 172$

$580 - 172$

$\boxed{408}$

The net profit for Sherlock was $408.

26) Find the result when the square of six is diminished by the product of five and four.

$6^2 - (5)(4)$

$36 - 20$

$\boxed{16}$

The result is 16.

28) Find the result when five times the difference of nine and three is decreased by ten.

$5(9-3) - 10$

$5(6) - 10$

$30 - 10$

$\boxed{20}$

The result is 20.

30) Find the result when the square of seven is increased by the cube of negative two.

$7^2 + (-2)^3$

$49 + (-8)$

$\boxed{-41}$

The result is $\boxed{-41}$

) –31   5) 92   8) 9   11) 3   14) 7y – 9 = 53

19) 3x + 8 = 5x – 6   23) $61   25) $3200   27) 60   29) 62   31) 45

2.1 Answers

1) $3 + 5 + 6 = 8 + 6$ Associative +	25) $2x + 3 + 7x + 8$ $\boxed{9x + 11}$
4) $9 + (-9) = 0$ Inverse +	28) $4 - 2w - 3w - 5$ $\boxed{-5w - 1}$
7) $\frac{7x}{7} = 1x$ Inverse ·	31) $8 + 3(2x - 6) + 7$ $8 + 6x - 18 + 7$ $\boxed{6x - 3}$
Use the indicated property to write a	
10) By commutative + $7 + 3 =$ $3 + 7$	34) $-2x - 3x - 4x - 5$ $\boxed{-9x - 5}$
13) By distributive property $6(x + 4) =$ $6x + 24$	
16) By associative + $2 + (4 + 7) = (2 + 4) + 7$	37) $y + 4(2y - 8) + 6$ $y + 8y - 32 + 6$ $\boxed{9y - 26}$
Use the distributive property and simp.	
19) $5(2a + 4) =$ $\boxed{10a + 20}$	40) $7(x + 5) + 2(x - 6)$ $7x + 35 + 2x - 12$ $\boxed{9x + 23}$
22) $3 + 4(2x + 5)$ $3 + 8x + 20$ $\boxed{8x + 23}$	

2) Distributive   5) Associative ·   8) Identity ·   11) -4 14) 1 17) $6x - 15$

20) $-15w + 12$ 23) $8y - 23$ 26) $6a + 6$ 29) $2k - 10$ 32) $8a - 14$ 35) $16x - 15$

38) $14x - 6$ 41) $-12a - 3$

1) $2a + 5 = 11$

$2a + 5 - 5 = 11 - 5$

$\dfrac{2a}{2} = \dfrac{6}{2}$

$\boxed{a = 3}$

Check: $2(3) + 5 \stackrel{?}{=} 11$
$6 + 5$
$\boxed{11 = 11}$

4) $-4x - 3 = 5$

$-4x - 3 + 3 = 5 + 3$

$\dfrac{-4x}{-4} = \dfrac{8}{-4}$

$\boxed{x = -2}$

Check: $-4(-2) - 3 \stackrel{?}{=} 5$
$8 - 3$
$\boxed{5 = 5}$

7) $-m + 5 = 18$

$-m + 5 - 5 = 18 - 5$

$-m = 13$

$-1(-m) = (-1)13$

$\boxed{m = -13}$

Check: $-(-13) + 5 \stackrel{?}{=} 18$
$13 + 5$
$\boxed{18 = 18}$

10) $7w - 9 = -23$

$7w - 9 + 9 = -23 + 9$

$\dfrac{7w}{7} = \dfrac{-14}{7}$

$\boxed{w = -2}$

Check: $7(-2) - 9 \stackrel{?}{=} -23$
$-14 - 9$
$\boxed{-23 = -23}$

13) $\dfrac{x}{6} - 4 = 1$

$\dfrac{x}{6} - 4 + 4 = 1 + 4$

$\dfrac{x}{6} = 5$

$\dfrac{x}{6}(6) = 5(6)$

$\boxed{x = 30}$

Check: $\dfrac{(30)}{6} - 4 \stackrel{?}{=} 1$
$5 - 4$
$\boxed{1 = 1}$

16) $8 - (-3x) = 59$

$8 + 3x - 8 = 59 - 8$

$\dfrac{3x}{3} = \dfrac{51}{3}$

$\boxed{x = 17}$

Check: $8 - [-3(17)] \stackrel{?}{=} 59$
$8 - (-51)$
$\boxed{59 = 59}$

2) $b = 1$ 5) $y = -1$ 8) $x = 12$ 11) $y = -5$ 14) $k = 10$ 17) $x = 8$

1) $3a + 2 = 14$

$3a + 2 - 2 = 14 - 2$

$\dfrac{3a}{3} = \dfrac{12}{3}$

$\boxed{a = 4}$

Check: $3(4) + 2 \stackrel{?}{=} 14$
$12 + 2$
$\boxed{14 = 14}$

4) $3 + \dfrac{n}{2} = 9$

$3 + \dfrac{n}{2} - 3 = 9 - 3$

$\dfrac{n}{2} = 6$

$\dfrac{n}{2}(2) = 6(2)$

$\boxed{n = 12}$

Check: $3 + \dfrac{12}{2} \stackrel{?}{=} 9$
$3 + 6$
$\boxed{9 = 9}$

7) $25 = -3 + 4w$

$25 + 3 = -3 + 4w + 3$

$\dfrac{28}{4} = \dfrac{4w}{4}$

$7 = w$

or $\boxed{w = 7}$

Check: $25 \stackrel{?}{=} -3 + 4(7)$
$-3 + 28$
$\boxed{25 = 25}$

10) Five more than twice a number is 37. Find the number.

Let x = the number

$5 + 2x = 37$

$5 + 2x - 5 = 37 - 5$

$\dfrac{2x}{2} = \dfrac{32}{2}$

$\boxed{x = 16}$

$\boxed{\text{The number is 16.}}$

12) Eight times a certain number increased by 30 will be 6. What is the number?

Let n = the number.

$8n + 30 = 6$

$8n + 30 - 30 = 6 - 30$

$\dfrac{8n}{8} = \dfrac{-24}{8}$

$\boxed{n = -3}$

$\boxed{\text{The number is } -3.}$

14) When a certain number divided by six was increased by 3 the result was ten. What was the number?

Let a = the number

$\dfrac{a}{6} + 3 = 10$

$\dfrac{a}{6} + 3 - 3 = 10 - 3$

$(6)\dfrac{a}{6} = 7(6)$

$\boxed{a = 42}$

$\boxed{\text{The number is 42.}}$

2) v = 13   5) a = 7   8) x = 2   11) 8   13) 4   15) 28

1) $5x - 2 + 2x = 40$

$7x - 2 + 2 = 40 + 2$

$\dfrac{7x}{7} = \dfrac{42}{7}$

$\boxed{x = 6}$

Check:
$5(6) - 2 + 2(6) \stackrel{?}{=} 40$
$30 - 2 + 12$
$\boxed{40 = 40}$

4) $6r - 2 = 2r - 26$

$6r - 2 - 2r = 2r - 26 - 2r$

$4r - 2 + 2 = -26 + 2$

$\dfrac{4r}{4} = \dfrac{-24}{4}$

$\boxed{r = -6}$

Check:
$6(-6) - 2 \stackrel{?}{=} 2(-6) - 26$
$-36 - 2 \quad\quad -12 - 26$
$\boxed{-38 = -38}$

7) $9 - 3w = 5w + 57$

$9 - 3w - 5w = 5w + 57 - 5w$

$-8w + 9 - 9 = 57 - 9$

$\dfrac{-8w}{-8} = \dfrac{48}{-8}$

$\boxed{w = -6}$

Check:
$9 - 3(-6) \stackrel{?}{=} 5(-6) + 57$
$9 + 18 \quad\quad -30 + 57$
$\boxed{27 = 27}$

10) $14 = 2x - 21 + 5x$

$14 + 21 = 7x - 21 + 21$

$\dfrac{35}{7} = \dfrac{7x}{7}$

$\boxed{x = 5}$

Check:
$14 \stackrel{?}{=} 2(5) - 21 + 5(5)$
$\quad\quad 10 - 21 + 25$
$\quad\quad -11 + 25$
$\boxed{14 = 14}$

Solve the following word problems:

13) Five times a certain number decreased by three is the same as twice the number increased by 18. Find the number.

Let x = the number

$5x - 3 = 2x + 18$

$5x - 3 - 2x = 2x + 18 - 2x$

$3x - 3 + 3 = 18 + 3$

$\dfrac{3x}{3} = \dfrac{21}{3}$

$\boxed{x = 7}$

$\boxed{\text{The number is 7.}}$

15) Four more than six times a number is twice the number added to 32. What is the number?

Let n = the number

$4 + 6n = 2n + 32$

$4 + 6n - 2n = 2n + 32 - 2n$

$4n + 4 - 4 = 32 - 4$

$\dfrac{4n}{4} = \dfrac{28}{4}$

$\boxed{n = 7}$

$\boxed{\text{The number is 7.}}$

2) $a = 12$ 5) $n = 7$ 8) $n = -7$ 11) $x = -4$ 14) -6 16) -7

1) $3(x+3) = 8(x-2)$

$3x + 9 - 8x = 8x - 16 - 8x$

$-5x + 9 - 9 = -16 - 9$

$\dfrac{-5x}{-5} = \dfrac{-25}{-5}$

$\boxed{x = 5}$

$3[(5)+3] = 8[(5)-2]$

Check: $3(8) \quad 8(3)$

$\boxed{24 = 24}$

4) $16 - 3(2b + 5) = 3b - 17$

$16 - 6b - 15 = 3b - 17$

$1 - 6b - 3b = 3b - 17 - 3b$

$-9b + 1 - 1 = -17 - 1$

$\dfrac{-9b}{-9} = \dfrac{-18}{-9}$

$\boxed{b = 2}$

7) $7y + 3(5y - 2) = 38$

$7y + 15y - 6 = 38$

$22y - 6 + 6 = 38 + 6$

$\dfrac{22y}{22} = \dfrac{44}{22}$

$\boxed{y = 2}$

20) Eight more than five times the sum of a number and six is 58 increased by three times the number. Find the number.

Let $x =$ the number

$8 + 5(x + 6) = 58 + 3x$

$8 + 5x + 30 = 58 + 3x$

$5x + 38 - 3x = 58 + 3x - 3x$

$2x + 38 - 38 = 58 - 38$

$2x = 20$

$\boxed{x = 10}$

$\boxed{\text{The number is 10.}}$

22) What number increased by four times the difference of twice the number and seven is the same as two added to five times the sum of the number and two?

Let $a =$ the number

$a + 4(2a - 7) = 2 + 5(a + 2)$

$a + 8a - 28 = 2 + 5a + 10$

$9a - 28 - 5a = 5a + 12 - 5a$

$4a - 28 + 28 = 12 + 28$

$\dfrac{4a}{4} = \dfrac{40}{4}$

$\boxed{a = 10}$

$\boxed{\text{The number is 10.}}$

2) $x = 3$ 5) $x = 10$ 8) $x = -1$ 11) $3 + 4(x - 10) = 2x + 58$ 13) $7(x - 8) - 5 = 86$

21) 11 23) 6

2.6 Answers

1) $5 + 2a = 17$
$5 + 2a - 5 = 17 - 5$
$\dfrac{2a}{2} = \dfrac{12}{2}$
$\boxed{a = 6}$

3) $4 - 6(2x+3) = 34$
$4 - 12x - 18 = 34$
$-12x - 14 + 14 = 34 + 14$
$\dfrac{-12x}{-12} = \dfrac{48}{-12}$
$\boxed{x = -4}$

5) $3m - 4(2m-7) = m - 44$
$3m - 8m + 28 = m - 44$
$-5m + 28 - m = m - 44 - m$
$-6m + 28 - 28 = -44 - 28$
$\dfrac{-6m}{-6} = \dfrac{-72}{-6}$
$\boxed{m = 12}$

Evaluate each of the following if $a = -2$

7) $a + b[w - x]$
$(-2) + (3)[(6) - (-5)]$
$-2 + 3[6 + 5]$
$-2 + 3[11]$
$-2 + 33$
$\boxed{31}$

10) Six more than twice a number is five times the sum of the number and three. Find the number.

Let x = the number
$6 + 2x = 5(x+3)$
$6 + 2x = 5x + 15$
$6 + 2x - 5x = 5x + 15 - 5x$
$-3x + 6 - 6 = 15 - 6$
$\dfrac{-3x}{-3} = \dfrac{9}{-3}$
$\boxed{x = -3}$

The number is -3.

12) If seven times a certain number is diminished by twice the difference of the number and six the result is twice the number decreased by three. Find the number.

Let a = the number
$7a - 2(a-6) = 2a - 3$
$7a - 2a + 12 = 2a - 3$
$5a + 12 - 2a = 2a - 3 - 2a$
$3a + 12 - 12 = -3 - 12$
$\dfrac{3a}{3} = \dfrac{-15}{3}$
$\boxed{a = -5}$

The number is -5.

14) Three less than six times the sum of a number and five will be nine. Find the number.

Let y = the number
$6(y+5) - 3 = 9$
$6y + 30 - 3 = 9$
$6y + 27 - 27 = 9 - 27$
$\dfrac{6y}{6} = \dfrac{-18}{6}$
$\boxed{y = -3}$

The number is -3.

16) Five times the difference of three times a number and seven is one more than six times the sum of twice a number and five. What is the number?

Let n = the number
$5(3n - 7) = 1 + 6(2n + 5)$
$15n - 35 = 1 + 12n + 30$
$15n - 35 - 12n = 12n + 31 - 12n$
$3n - 35 + 35 = 31 + 35$
$\dfrac{3n}{3} = \dfrac{66}{3}$
$\boxed{n = 22}$

The number is 22.

2) $x = 6$ **4)** $x = 3$ **6)** $y = 11$ **8)** -3 **11)** 13   **13)** 33

15) 3   **17)** 8

3.1 Answers

1. The larger of two numbers is five more than three times the smaller number. If their sum is 29, find the numbers.

x = smaller number
$3x+5$ = larger number

$x + 3x+5 = 29$
$4x + 5 - 5 = 29 - 5$
$\dfrac{4x}{4} = \dfrac{24}{4}$

$\boxed{x = 6}$

$3(6) + 5 =$
$18 + 5 = \boxed{23}$

The numbers are 6 and 23.

5. In a certain class, there are eight more than twice as many women as men. If the class has 47 students, how many women are in the class?

x = number of men
$2x+8$ = " " women

$x + 2x + 8 = 47$
$3x + 8 - 8 = 47 - 8$
$\dfrac{3x}{3} = \dfrac{39}{3}$

$\boxed{x = 13}$

$2(13) + 8 = \boxed{34}$

There were 34 women in the class.

9. The larger of two numbers is five less than times the smaller. Seven times the larger diminished by twice the smaller is 41. Find larger number.

x = smaller number
$3x-5$ = larger "

$7(3x-5) - 2x = 41$
$21x - 35 - 2x = 41$
$19x - 35 + 35 = 41 + 35$
$19x = 76$

$\boxed{x = 4}$

$3(4) - 5 = \boxed{7}$

The larger number is 7.

3. Marta recently paid $290 for a bicycle and a helmet. She paid two dollars more than five times as much for the bike as the helmet. What was the price of the bike?

x = price of helmet
$5x+2$ = " " bike

$x + 5x + 2 = 290$
$6x + 2 - 2 = 290 - 2$
$\dfrac{6x}{6} = \dfrac{288}{6}$

$\boxed{x = 48}$

$5(48) + 2 =$
$240 + 2 = \boxed{242}$

The price of bike was $242.

7. The weight of Tarzan is 30 pounds less than twice the weight of Madonna. Four times Madonna's weight added to three times Tarzan's weight is 960 pounds. Find the weight of Ferd Burfil who weighs five pounds less than Tarzan.

$2x-30$ = weight of Tarzan
x = " " Madonna

$4x + 3(2x-30) = 960$
$4x + 6x - 90 = 960$
$10x - 90 + 90 = 960 + 90$
$\dfrac{10x}{10} = \dfrac{1050}{10}$

$\boxed{x = 105}$ MADONNA

$2(105) - 30 = \boxed{180}$ TARZAN

Ford weighs 175 pounds.

11. Mortimer is three years older than twice the age of Alfonso. If the sum of their ages is 51, find the age of Mortimer.

a = age of Alfonso
$2a+3$ = " " Mortimer

$a + 2a + 3 = 51$
$3a + 3 - 3 = 51 - 3$
$\dfrac{3a}{3} = \dfrac{48}{3}$

$\boxed{a = 16}$

$2(16) + 3 = \boxed{35}$

Mortimer is 35 years old.

2) 37   4) $6   6) 5 and 10   8) 17   10) 15 miles   12) $708

1. A triangle has a perimeter of 78 meters. Two sides have the same length and the third side is six meters less than the sum of the other two sides. Find the length of the largest side.

s = length of 1st side of △
s =   "   " 2nd " "
$2s-6$ = " " 3rd " "

$s + s + 2s - 6 = 78$
$4s - 6 + 6 = 78 + 6$
$\dfrac{4s}{4} = \dfrac{84}{4}$
$\boxed{s = 21}$
$2(21) - 6 = \boxed{36}$

$\boxed{\text{The largest side is 36 m. long.}}$

5. The floor of Batman's costume storage closet is nine feet more than three times its width. If the perimeter is 138 feet, what is its length?

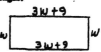

w = width of closet
$3w + 9$ = length " "

$2\ell + 2w = P$ FORMULA
$2(3w+9) + 2w = 138$
$6w + 18 + 2w = 138$
$8w + 18 - 18 = 138 - 18$
$8w = 120$
$\boxed{w = 15}$
$3(15) + 9 = \boxed{54}$

$\boxed{\text{The length of the closet is 54 ft.}}$

3. Carol recently bought a book, a computer disk, and a pogo stick, paying a total of $126. She paid two dollars more than three times as much for the book as the disk, while the pogo stick cost twice as much as the book. Find the cost of the pogo stick.

$3x+2$ = cost of book
x = "   " disk
$2(3x+2)$ = " " Pogo stick

$3x + 2 + x + 2(3x+2) = 126$
$4x + 2 + 6x + 4 = 126$
$10x + 6 - 6 = 126 - 6$
$\dfrac{10x}{10} = \dfrac{120}{10}$
$\boxed{x = 12}$
$2[3(12)+2] =$
$2[38] = \boxed{76}$

$\boxed{\text{The pogo stick cost \$76.}}$

7. Ivan slept two hours less than Carlos, while Boris slept as much as Ivan and Carlos combined. If these guys slept a total of 32 hours, how many hours did Boris sleep?

$2x-2$ = number of hours Boris slept
x =   "   "   "   Carlos "
$x-2$ = "   " "   Ivan "

$2x - 2 + x + x - 2 = 32$
$4x - 4 + 4 = 32 + 4$
$\dfrac{4x}{4} = \dfrac{36}{4}$
$\boxed{x = 9}$
$2(9) - 2 = \boxed{16}$

$\boxed{\text{Boris slept 16 hours.}}$

2) 25 cm.   4) 34 problems   6) 20 cm. by 14 cm.   8) 30

3.3 Answers

1) A computer and printer can be purchased in the Book Store for $1025. Fannie Farkel decided to buy five printers and two computers paying a total of $3175. **Find the cost of each.**

x = cost of computer
$1025-x$ = " " printer

$5(1025-x) + 2x = 3175$
$5125 - 5x + 2x = 3175$
$5125 - 3x - 5125 = 3175 - 5125$
$\dfrac{-3x}{-3} = \dfrac{-1950}{-3}$
$\boxed{x = 650}$
$1025 - 650 = \boxed{375}$

The computer cost $650.
The printer cost $375.

3) The age of Pat is three times the age of Ben, while Etar is five years younger than Pat. If the sum of their ages is 79, **find the age of Etar.**

$3a$ = age of Pat
a = " " Ben
$3a-5$ = " " Etar

$3a + a + 3a - 5 = 79$
$7a - 5 + 5 = 79 + 5$
$\dfrac{7a}{7} = \dfrac{84}{7}$
$\boxed{a = 12}$
$3(12) - 5 = \boxed{31}$

Etar is 31 years old.

9) $6 + 2(3x-5) = 10x - 8$
$6 + 6x - 10 = 10x - 8$
$6x - 4 - 10x = 10x - 8 - 10x$
$-4x - 4 + 4 = -8 + 4$
$\dfrac{-4x}{-4} = \dfrac{-4}{-4}$
$\boxed{x = 1}$

Check:
$6 + 2[3(1) - 5] \stackrel{?}{=} 10(1) - 8$
$6 + 2[-2] \stackrel{?}{=} 10 - 8$
$6 + (-4) \stackrel{?}{=} 2$
$\boxed{2 = 2}$

12) $-4 - 4(2x+5) + 5x = x$
$-4 - 8x - 20 + 5x = x$
$-3x - 24 + 3x = x + 3x$
$\dfrac{-24}{4} = \dfrac{4x}{4}$
$\boxed{x = -6}$

$-4 - 4[2(-6) + 5] + 5(-6) \stackrel{?}{=} -6$
Check:
$-4 - 4[-12+5] + (-30) \stackrel{?}{=} -6$
$-4 - 4[-7] - 30 \stackrel{?}{=} -6$
$-4 + 28 - 30 \stackrel{?}{=} -6$
$\boxed{-6 = -6}$

Evaluate each of the following if

15) $a^2n - x^2v$
$(-3)^2(2) - (-5)^2(8)$
$9(2) - (25)(8)$
$18 - 200$
$\boxed{-182}$

5) Ferd Burfil has a rectangular back yard. He wo[uld] like to build a fence on the two sides and the bac[k]. The back is 12 feet less than twice the length of [each] side. The total length of the fence is 156 feet. [How] long is the back fence?

BACK: $2x-12$
x
House

x = length of each side fence
$2x-12$ = " " back fence

$x + x + 2x - 12 = 156$
$4x - 12 + 12 = 156 + 12$
$\dfrac{4x}{4} = \dfrac{168}{4}$
$\boxed{x = 42}$
$2(42) - 12 = \boxed{72}$

The back fence is 72 feet long.

7) A sack has 100 marbles, some are red, some are blue. Four times the number of red marbles diminished by twice the number of blue marbles i[s] 286. **How many of each color marble are in the sack?**

x = number of red marbles
$100-x$ = " " blue "

$4x - 2(100-x) = 286$
$4x - 200 + 2x = 286$
$6x - 200 + 200 = 286 + 200$
$\dfrac{6x}{6} = \dfrac{486}{6}$
$\boxed{x = 81}$
$100 - 81 = \boxed{19}$

There are 81 red marbles and 19 blue marbles.

---

2) 7   4) $7   6) 15 boys   8) $507   10) x = 4   13) a = –3   16) 40

1. Kareem bought a Mickey Mouse pencil and a blue cap with a propeller on top, paying a total of $19. The cap cost four dollars more than twice the cost of the pencil. Find the cost of the cap.

c = cost of pencil
$2c+4$ = "   " cap

$$c + 2c + 4 = 19$$
$$3c + 4 - 4 = 19 - 4$$
$$\frac{3c}{3} = \frac{15}{3}$$
$$\boxed{c = 5}$$
$$2(5) + 4 = \boxed{14}$$

The cap cost $14.

5. The combined age of Amanda and her Mom is 71 years. Her Mom is seven years more than three times the age of Amanda. How old is the Mother?

a = age of Amanda
$3a+7$ = age of Mother

$$a + 3a + 7 = 71$$
$$4a + 7 - 7 = 71 - 7$$
$$\frac{4a}{4} = \frac{64}{4}$$
$$\boxed{a = 16}$$
$$3(16) + 7 = 55$$

Mother is 55 years old.

3. The lowest grade on a recent math exam was a grade of C. (Students learned to think like a mathematician.) There were twice as many A grades as B grades and the number of C grades was nine less than the number of A grades. How many grades were A if the class had 36 students?

$2x$ = number of A grades
x = "   " B "
$2x-9$ = "   " C "

$$2x + x + 2x - 9 = 36$$
$$5x - 9 + 9 = 36 + 9$$
$$\frac{5x}{5} = \frac{45}{5}$$
$$\boxed{x = 9}$$
$$2(9) = \boxed{18}$$

There were 18 A grades.

7. The sum of two numbers is 63. When twice the smaller is added to five times the larger the result is 276. Find the two numbers.

s = smaller number
$63 - s$ = larger   "

$$2s + 5(63-s) = 276$$
$$2s + 315 - 5s = 276$$
$$-3s + 315 - 315 = 276 - 315$$
$$\frac{-3s}{-3} = \frac{-39}{-3}$$
$$\boxed{s = 13}$$
$$63 - 13 = \boxed{50}$$

The two numbers are 13 and 50.

2) 27 feet   4) $13   6) 49 meters   8) 149 pounds

3.5 Answers

1) $5 + 2a = 47$

$5 + 2a - 5 = 47 - 5$

$\dfrac{2a}{2} = \dfrac{42}{2}$

$\boxed{a = 21}$

4) $9 + 5(2x - 4) = 4 + 3(2x - 9)$

$9 + 10x - 20 = 4 + 6x - 27$

$10x - 11 - 6x = 6x - 23 - 6x$

$4x - 11 + 11 = -23 + 11$

$\dfrac{4x}{4} = \dfrac{-12}{4}$

$\boxed{x = -3}$

Evaluate each of the following if a =

7) $a + b[x - w]$

$(-2) + (3)[(-5) - (4)]$

$-2 + 3[-9]$

$-2 - 27$

$\boxed{-29}$

10. The combined weight of Boris and Ivan is 360 pounds. If Boris weighs six pounds more than twice the weight of Ivan, how much does Boris weigh?

$2x + 6$ = weight of Boris
x = " " Ivan

$2x + 6 + x = 360$

$3x + 6 - 6 = 360 - 6$

$\dfrac{3x}{3} = \dfrac{354}{3}$

$\boxed{x = 118}$

$2(118) + 6 = \boxed{242}$

Boris weighs 242 pounds.

12. Tarzan is eight years more than three times the age of Ferd Burfill. If the sum of their ages is 84, <u>find the age of Batman who is five years younger than Tarzan.</u>

x = age of Ferd
$3x + 8$ = " " Tarzan

$x + 3x + 8 = 84$

$4x + 8 - 8 = 84 - 8$

$\dfrac{4x}{4} = \dfrac{76}{4}$

$\boxed{x = 19}$ Ferd

$3(19) + 8 = \boxed{65}$ Tarzan

Batman is 60 yr. old.

14. When Steve Martin was 38 years old he bought a rectangular lot at a total cost of $240,000. The lot was four feet less than three times its width. If the perimeter was 432 feet, <u>what was the length.</u>

x = width of lot
$3x - 4$ = length " "

Perimeter Formula $2l + 2w = P$

$2(3x - 4) + 2x = 432$

$6x - 8 + 2x = 432$

$8x - 8 + 8 = 432 + 8$

$\dfrac{8x}{8} = \dfrac{440}{8}$

$\boxed{x = 55}$

$3(55) - 4 = \boxed{161}$

The length of the lot was 161 feet.

16. Wanda is three years less than five times as old as Gilda. If twice the age of Wanda is decreased by four times the age of Gilda, the result would be exactly 78 years. How old will Wanda be in two short years?

$5a - 3$ = age of Wanda
a = " " Gilda

$2(5a - 3) - 4a = 78$

$10a - 6 - 4a = 78$

$6a - 6 + 6 = 78 + 6$

$\dfrac{6a}{6} = \dfrac{84}{6}$

$\boxed{a = 14}$

$5(14) - 3 = \boxed{67}$

Wanda will be 69 in 2 short years.

2) $x = 17$ **5)** $m = 3$ **8)** -7 **11)** 80   **13)** 26 and 47   **15)** $90

17) 10 Tacos

4.1 Answers

1. $\frac{3}{8}$

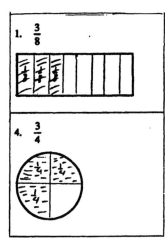

4. $\frac{3}{4}$

16. $63 = 3 \cdot 21$
$= 3 \cdot 3 \cdot 7$

The prime factors of 63 are 3, 3, 7

Reduce each of the following using pri

19. $\frac{12}{15} = \frac{2 \cdot 2 \cdot 3}{3 \cdot 5}$
$= \boxed{\frac{4}{5}}$

Find the GCF (Greatest Common Factor)

7. 6 and 8
$6 = 2 \cdot 3$
$8 = 2 \cdot 2 \cdot 2$
$\boxed{GCF = 2}$

22. $\frac{42}{35} = \frac{2 \cdot 3 \cdot 7}{5 \cdot 7}$
$= \boxed{\frac{14}{5}}$

Reduce each of the following using GCF

10. $\frac{20}{24} = \frac{2 \cdot 2 \cdot 5}{2 \cdot 2 \cdot 2 \cdot 3}$
$= \boxed{\frac{5}{6}}$

Build each of the following fractions to

25. $\frac{3}{4} \cdot \frac{6}{6} = \frac{18}{24}$

13. $\frac{16}{40} = \frac{2 \cdot 2 \cdot 2 \cdot 2}{2 \cdot 2 \cdot 2 \cdot 5}$
$= \boxed{\frac{2}{5}}$

28. $\frac{7}{3} \cdot \frac{3}{5} = \frac{21}{9}$

GCF = 4   11) $\frac{5}{6}$ 14) $\frac{5}{2}$ 17) 2, 2, 3, 3 and 5   20) $\frac{3}{4}$ 23) $\frac{11}{7}$

45   29) 64

1) $\dfrac{5}{7} \cdot \dfrac{2}{3} = \boxed{\dfrac{10}{21}}$

4) $\dfrac{7}{15} \cdot \dfrac{5}{7}$

$\dfrac{\cancel{7}}{3 \cdot \cancel{5}} \cdot \dfrac{\cancel{5}}{\cancel{7}}$

$\boxed{\dfrac{1}{3}}$

7) $\dfrac{5}{8} \cdot \dfrac{12}{7}$

$\dfrac{5}{\cancel{2} \cdot \cancel{2} \cdot 2} \cdot \dfrac{\cancel{2} \cdot \cancel{2} \cdot 3}{7}$

$\boxed{\dfrac{15}{14}}$

Find the quotient.

10) $\dfrac{2}{3} \div \dfrac{5}{7}$

$= \dfrac{2}{3} \cdot \dfrac{7}{5}$

$= \boxed{\dfrac{14}{15}}$

13) $\dfrac{4}{5} \div \dfrac{16}{3}$

$= \dfrac{\cancel{4}}{5} \cdot \dfrac{3}{\cancel{16}\,4}$

$= \boxed{\dfrac{3}{20}}$

16) $\dfrac{2}{3} + \dfrac{5}{3}$

$= \boxed{\dfrac{7}{3}}$

19) $\dfrac{5}{6} + \dfrac{11}{6}$

$= \dfrac{16}{6}$

$= \boxed{\dfrac{8}{3}}$

Find the difference.

22) $\dfrac{8}{9} - \dfrac{2}{9}$

$= \dfrac{6}{9}$

$= \boxed{\dfrac{2}{3}}$

25) $\dfrac{11}{6} - \dfrac{7}{6}$

$= \dfrac{4}{6}$

$= \boxed{\dfrac{2}{3}}$

28) $\dfrac{5}{6} \cdot \dfrac{2}{3}$

$= \dfrac{5}{\cancel{2} \cdot 3} \cdot \dfrac{\cancel{2}}{3}$

$= \boxed{\dfrac{5}{9}}$

31) $\dfrac{47}{48} - \dfrac{5}{6} \cdot \dfrac{7}{8}$

$= \dfrac{47}{48} - \dfrac{35}{48}$

$= \dfrac{12}{48} \quad = \dfrac{\sqrt{2 \cdot 1}}{\cancel{12} \cdot 3}$

$= \boxed{\dfrac{1}{3}}$

Use a "Factor tree" to factor each of th

34) $150 = 2 \cdot 75$

$= 2 \cdot 3 \cdot 25$

$= 2 \cdot 3 \cdot 5 \cdot 5$

The prime factors of 150 are 2, 3, 5 and 5

Find the GCF (greatest common facto

37) 30 and 45

$30 = 2 \cdot 3 \cdot 5$

$45 = 3 \cdot 3 \cdot 5$

$GCF = 3 \cdot 5 = 15$

---

2) $\dfrac{7}{20}$ 5) $\dfrac{3}{2}$ 8) $\dfrac{10}{3}$ 11) $\dfrac{1}{4}$ 14) $\dfrac{7}{16}$ 17) 1   20) $\dfrac{3}{5}$

23) $\dfrac{1}{6}$ 26) 1   29) $\dfrac{6}{7}$ 32) $\dfrac{50}{7}$

35) The prime factors of 240 are 2, 2, 2, 2, 3, and 5   38) GCF = 6

1) $\frac{3}{16}+\frac{7}{16}$

$= \frac{10}{16}$

$= \boxed{\frac{5}{8}}$

4) $\frac{7}{9}-\frac{2}{9}$

$\boxed{\frac{5}{9}}$

7) $\frac{3}{8}+\frac{5}{6}$

$\frac{3}{8}\cdot\frac{3}{3} + \frac{5}{6}\cdot\frac{4}{4}$

$\frac{9}{24} + \frac{20}{24}$

$\boxed{\frac{29}{24}}$

10) $\frac{5}{6}+\frac{2}{3}-\frac{1}{2}$

$=\frac{5}{6} + \frac{2}{3}\cdot\frac{2}{2} - \frac{1}{2}\cdot\frac{3}{3}$

$=\frac{5}{6} + \frac{4}{6} - \frac{3}{6}$

$= \frac{6}{6}$

$= \boxed{1}$

13) $\frac{5}{6}\cdot\frac{2}{3}$

$= \boxed{\frac{5}{9}}$

16) $\frac{3}{4}+\frac{2}{3}\cdot\frac{3}{5}$

$= \frac{3}{4}+ \frac{2}{5}$

$= \frac{3}{4}\cdot\frac{5}{5} + \frac{2}{5}\cdot\frac{4}{4}$

$= \frac{15}{20} + \frac{8}{20}$

$= \boxed{\frac{23}{20}}$

19) $\frac{9}{2}-\frac{2}{3}+\frac{3}{4}$

$= \frac{9}{2}\cdot\frac{6}{6} - \frac{2}{3}\cdot\frac{4}{4} + \frac{3}{4}\cdot\frac{3}{3}$

$= \frac{54}{12} - \frac{8}{12} + \frac{9}{12}$

$= \frac{46}{12} + \frac{9}{12}$

$= \boxed{\frac{55}{12}}$

22) $\frac{25}{4}\cdot\frac{15}{8}+\frac{5}{6}$

$= \frac{25}{4}\cdot\frac{8^2}{15_3} + \frac{5}{6}$

$= \frac{10}{3} + \frac{5}{6}$

$= \frac{10}{3}\cdot\frac{2}{2} + \frac{5}{6}$

$= \frac{20}{6} + \frac{5}{6}$

$= \boxed{\frac{25}{6}}$

2) $\frac{3}{4}$ 5) $\frac{9}{8}$ 8) $\frac{5}{12}$ 11) $\frac{11}{12}$ 14) $\frac{7}{10}$ 17) $\frac{9}{5}$

20) $\frac{17}{30}$ 23) $\frac{7}{6}$

4.4 Answers

1) $\frac{8}{3} = \boxed{2\frac{2}{3}}$

Express each of the following mixed f

4) $3\frac{2}{5} = \boxed{\frac{17}{5}}$

Perform the indicated operation:

7) $7\frac{2}{3} + 5\frac{5}{6}$
$= 7\frac{2 \cdot 2}{3 \cdot 2} + 5\frac{5}{6}$
$= 7\frac{4}{6} + 5\frac{5}{6}$
$= 12\frac{9}{6}$
$= 12\frac{3}{2} = \boxed{13\frac{1}{2}}$

10) $2\frac{1}{3} \cdot 1\frac{3}{14}$
$= \frac{7}{3} \cdot \frac{17}{14}_2$
$= \boxed{\frac{17}{6}} \text{ or } \boxed{2\frac{5}{6}}$

13) $3\frac{5}{6} + 2\frac{2}{3}$
$= 3\frac{5}{6} + 2\frac{2}{3} \cdot \frac{2}{2}$
$= 3\frac{5}{6} + 2\frac{4}{6}$
$= 5\frac{9}{6}$
$= 5\frac{3}{2} = \boxed{6\frac{1}{2}}$

16) $8\frac{13}{23} - 3\frac{5}{6}$
$= 8\frac{3}{6} - 3\frac{5}{6}$
$= 7\frac{9}{6} - 3\frac{5}{6}$
$= 4\frac{4}{6}$
$= \boxed{4\frac{2}{3}}$

19) $3\frac{1}{2} \cdot \frac{3}{8} \div \frac{5}{16}$
$= \frac{7}{2} \cdot \frac{3}{8} \div \frac{5}{16}$
$= \frac{21}{16} \cdot \frac{16}{5}$
$= \boxed{\frac{21}{5}} \text{ or } \boxed{4\frac{1}{5}}$

22) $5\frac{2}{3} \div 4\frac{2}{9} + 3\frac{1}{2}$
$= \frac{17}{3} \div \frac{38}{9} + 3\frac{1}{2}$
$= \frac{17}{3} \cdot \frac{9^3}{38} + 3\frac{1}{2}$
$= \frac{51}{38} + \frac{7}{2} \cdot \frac{19}{19}$
$= \frac{51}{38} + \frac{133}{38}$
$= \frac{184}{38}$
$= \frac{92 \cdot 2}{19 \cdot 2} = \boxed{\frac{92}{19}} \text{ or } \boxed{4\frac{16}{19}}$

25) $\left[\frac{5}{4}\right]^2 + 3\frac{1}{2} \cdot \frac{19}{56}$
$= \frac{25}{16} + \frac{7}{2} \cdot \frac{19}{56}_8$
$= \frac{25}{16} + \frac{19}{16}$
$= \frac{44}{16}$
$= \boxed{\frac{11}{4}} \text{ or } \boxed{2\frac{3}{4}}$

28) For Mother's Day, Kent baked a cake which called for $2\frac{1}{4}$ cups of flour and $1\frac{1}{2}$ cups of sugar. Because everyone seemed to be on a diet, he decided to use only $\frac{2}{3}$ of the recommended ingredients. How much flour should he use?

$2\frac{1}{4} \cdot \frac{2}{3}$
$= \frac{9}{4}^3 \cdot \frac{2}{3}$
$= \frac{3}{2} \text{ or } 1\frac{1}{2}$

Kent should use $1\frac{1}{2}$ cup flour.

30) Fannie Farkel needed $12\frac{4}{5}$ yards of material to make four table cloths. How much material did she use on each table cloth?

$12\frac{4}{5} \div 4$
$= \frac{64}{5}^{16} \cdot \frac{1}{4}$
$= \frac{16}{5} \text{ or } 3\frac{1}{5}$

Fannie used $3\frac{1}{5}$ yards on each table cloth.

32) Alex uses $2\frac{3}{8}$ feet of ribbon on each Christmas wreath that he makes. If he made 28 wreaths, how much ribbon did he use?

$2\frac{3}{8} \cdot 28$
$= \frac{19}{8} \cdot 28^7$
$= \frac{133}{2} \text{ or } 66\frac{1}{2}$

Alex used $66\frac{1}{2}$ feet of ribbon.

34) Michael Barisnakov bought 120 yards of a material to make special dance outfits for his new tour. Each of the 24 outfits required $3\frac{2}{3}$ yards of material. If Delta Burke bought all of the remaining material at $14 per yard, how much did it cost her?

$24 \cdot 3\frac{2}{3}$
$= 24 \cdot \frac{11}{3}$
$= 88$ yds used
$120 - 88 = 32$ yds remained
$(32)(14) = \$448$

Delta paid $448 for the remaining 32 yds.

2) $2\frac{2}{5}$ 5) $\frac{19}{3}$ 8) $7\frac{1}{8}$ 11) $\frac{26}{3}$ or $8\frac{2}{3}$ 14) $5\frac{1}{4}$ 17) 30

20) $10\frac{1}{4}$ 23) $11\frac{23}{75}$ 26) $3\frac{9}{10}$ 29) 27 years old 31) $7\frac{3}{8}$ pounds

33) $24 35) $30\frac{1}{3}$ miles

4.5 Answers

1) $6\frac{2}{3} + 5\frac{1}{2} - 2\frac{5}{6}$

$= 6\frac{2}{3} \cdot \frac{2}{2} + 5\frac{1}{2} \cdot \frac{3}{3} - 2\frac{5}{6}$

$= 6\frac{4}{6} + 5\frac{3}{6} - 2\frac{5}{6}$

$= 11\frac{7}{6} - 2\frac{5}{6}$

$= 9\frac{2}{6}$

$= \boxed{9\frac{1}{3}}$

4) $1\frac{2}{7} + 2\frac{3}{5} \cdot 1\frac{3}{7}$

$= 1\frac{2}{7} + \frac{13}{5} \cdot \frac{10}{7}^2$

$= 1\frac{2}{7} + \frac{26}{7}$

$= 1\frac{2}{7} + 3\frac{5}{7}$

$= 4\frac{7}{7}$

$= \boxed{5}$

7) $4\frac{3}{5} \cdot 1\frac{3}{7} \cdot 5\frac{1}{4}$

$= \frac{23}{5} \cdot \frac{10}{7}^2 \cdot \frac{21}{4}^3$

$= \boxed{\frac{69}{2}} \text{ or } \boxed{34\frac{1}{2}}$

10) $2\frac{2}{3} \div 1\frac{5}{6} \cdot 4\frac{2}{5}$

$= \frac{8}{3} \div \frac{11}{6} \cdot \frac{22}{5}$

$= \frac{8}{3} \cdot \frac{6}{11}^2 \cdot \frac{22}{5}^2$

$= \boxed{\frac{32}{5}} \text{ or } \boxed{6\frac{2}{5}}$

13) $1\frac{7}{12} + 8\frac{3}{4}\cdot\frac{3}{3} - 4\frac{5}{6}\cdot\frac{2}{2}$

$= 1\frac{7}{12} + 8\frac{9}{12} - 4\frac{10}{12}$

$= 9\frac{16}{12} - 4\frac{10}{12}$

$= 5\frac{6}{12}$

$= \boxed{5\frac{1}{2}}$

Solve the following problems.

16) Rocco paid exactly $5 for the ingredients for his famous salad. He used $2\frac{1}{4}$ pounds of tomatoes, $1\frac{2}{3}$ pounds of cucumbers, $4\frac{3}{4}$ pounds of lettuce and $\frac{2}{3}$ pounds of garlic. By 8 PM he and his six guests had eaten the entire salad. On the average how much did each person eat?

$2\frac{1}{4} + 1\frac{2}{3} + 4\frac{3}{4} + \frac{2}{3}$

$= 2\frac{1\cdot3}{4\cdot3} + 1\frac{2\cdot4}{3\cdot4} + 4\frac{3\cdot3}{4\cdot3} + \frac{2\cdot4}{3\cdot4}$

$= 2\frac{3}{12} + 1\frac{8}{12} + 4\frac{9}{12} + \frac{8}{12}$

$= 7\frac{28}{12}$

$= 7\frac{7}{3}$

$= 9\frac{1}{3}$ total salad eaten by 7 people

$9\frac{1}{3} \div 7$

$= \frac{28}{3} \cdot \frac{1}{7} = \frac{4}{3} = 1\frac{1}{3}$ — EACH person ate $1\frac{1}{3}$ pounds

18) Find the result when the square of $\frac{3}{5}$ is added to the product of $\frac{4}{5}$ and $\frac{2}{3}$.

$\left(\frac{3}{5}\right)^2 + \frac{4}{5} \cdot \frac{2}{3}$

$= \frac{9}{25}\cdot\frac{3}{3} + \frac{8}{15}\cdot\frac{5}{5}$

$= \frac{27}{75} + \frac{40}{75}$

$= \boxed{\frac{67}{75}}$

The result would be $\frac{67}{75}$.

20) Multiply $3\frac{1}{3}$ by the difference of $5\frac{2}{3}$ and $2\frac{1}{6}$

$3\frac{1}{3} \cdot \left[5\frac{2/2}{3/2} - 2\frac{1}{6}\right]$

$= \frac{10}{3}\left[5\frac{4}{6} - 2\frac{1}{6}\right]$

$= \frac{10}{3}\left[3\frac{3}{6}\right]$

$= \frac{10}{3}\left[3\frac{1}{2}\right]$

$= \frac{10}{3}^5 \cdot \frac{7}{2}$

$= \boxed{\frac{35}{3}} \text{ or } \boxed{11\frac{2}{3}}$

The result would be $11\frac{2}{3}$.

22) Subtract $2\frac{3}{4}$ from the product of $1\frac{4}{5}$ and $6\frac{2}{3}$.

$\left(1\frac{4}{5}\right)\cdot\left(6\frac{2}{3}\right) - 2\frac{3}{4}$

$= \frac{9}{5}^3 \cdot \frac{20}{3}^4 - 2\frac{3}{4}$

$= 12 - 2\frac{3}{4}$

$= 11\frac{4}{4} - 2\frac{3}{4}$

$= \boxed{9\frac{1}{4}}$

The result would be $9\frac{1}{4}$.

2) $9\frac{2}{3}$ **5)** $5\frac{1}{12}$ **8)** $5\frac{17}{24}$ **11)** $58\frac{3}{4}$ **14)** $5\frac{5}{16}$ **17)** $2\frac{1}{8}$ pounds

19) $3\frac{2}{3}$ **21)** $8\frac{7}{8}$ **23)** $\frac{15}{16}$ **25)** $4\frac{15}{16}$

4.6 Answers

1) $5\frac{3}{4} + 2\frac{5}{6} - 3\frac{2}{3}$

$= 5\frac{3}{4}\left(\frac{3}{3}\right) + 2\frac{5}{6}\left(\frac{2}{2}\right) - 3\frac{2}{3}\left(\frac{4}{4}\right)$

$= 5\frac{9}{12} + 2\frac{10}{12} - 3\frac{8}{12}$

$= 7\frac{19}{12} - 3\frac{8}{12}$

$= \boxed{4\frac{11}{12}}$

4) $6\frac{1}{4} \div 1\frac{7}{8} + 2\frac{1}{7} \cdot \frac{7}{10}$

$= \frac{25}{4} \div \frac{15}{8} + \frac{15}{7} \cdot \frac{7}{10}$

$= \frac{25}{4} \cdot \frac{8}{15} + \frac{3}{2}$

$= \frac{10}{3} + \frac{3}{2}\left(\frac{3}{3}\right)$

$= \frac{20}{6} + \frac{9}{6}$

$= \boxed{\frac{29}{6}}$ or $\boxed{4\frac{5}{6}}$

7) $5\frac{1}{3}\left[12\frac{7}{8} - 5\frac{3}{4}\right]$

$= \frac{16}{3}\left[12\frac{7}{8} - 5\frac{6}{8}\right]$

$= \frac{16}{3} \cdot 7\frac{1}{8}$

$= \frac{16}{3} \cdot \frac{57}{8}$

$= \boxed{38}$

10) $ab + nw$

$\left(\frac{3}{4}\right)\left(\frac{2}{3}\right) + \left(\frac{7}{8}\right)\left(\frac{2}{10}\right)$

$\frac{1}{2} \cdot \frac{8}{8} + \frac{7}{16}$

$\frac{8}{16} + \frac{7}{16}$

$\boxed{\frac{15}{16}}$

12) $ab - ax$

$\left(\frac{3}{4}\right)\left(\frac{2}{3}\right) - \left(\frac{3}{8}\right)\left(\frac{5}{2}\right)$

$\frac{1}{2} \cdot \frac{4}{4} - \frac{15}{8}$

$\frac{4}{8} - \frac{15}{8}$

$\boxed{-\frac{11}{8}}$ or $\boxed{-1\frac{3}{8}}$

Solve each of the following problems:

14) Jung Han ran an average of 8 miles per hour for 2 hours and 40 minutes. How far did he run? (Use the distance formula: $d = r \cdot t$)

2 hrs and 40 min = $2\frac{40}{60}$ hrs = $2\frac{2}{3}$ hrs.

$d = rt$
$d = 8 \cdot 2\frac{2}{3}$
$d = 8 \cdot \frac{8}{3}$
$d = \frac{64}{3}$ or $21\frac{1}{3}$

Jung ran $21\frac{1}{3}$ miles

16) Nathan can ride his bike an average of 14 MPH. Gertraud can hop on her pogo stick an average of 2 MPH. If they started at the same spot at 2 PM and went in the same direction until 3:15 PM, how far ahead would Nathan be?

2PM to 3:15 PM would $1\frac{15}{60}$ hr
$1\frac{15}{60} = 1\frac{1}{4}$ hr or $\frac{5}{4}$ hr time

Nathan — use $d = rt$ — Gertraud
$d = 14 \cdot \frac{5}{4}$ $d = 2 \cdot \frac{5}{4}$
$d = \frac{35}{2}$ or $17\frac{1}{2}$ miles $d = \frac{5}{2}$ or $2\frac{1}{2}$ miles

$17\frac{1}{2} - 2\frac{1}{2} = 15$ miles

Nathan would be 15 miles ahead

18) Add $3\frac{5}{6}$ to the product of $2\frac{2}{3}$ and $1\frac{3}{4}$.

$3\frac{5}{6} + 2\frac{2}{3} \cdot 1\frac{3}{4}$

$= 3\frac{5}{6} + \frac{8}{3} \cdot \frac{7}{4}$

$= 3\frac{5}{6} + \frac{14}{3}$

$= 3\frac{5}{6} + 4\frac{2}{3} \cdot \frac{2}{2}$

$= 3\frac{5}{6} + 4\frac{4}{6}$

$= 7\frac{9}{6}$

$= 7\frac{3}{2} = \boxed{8\frac{1}{2}}$

The result is $8\frac{1}{2}$

20) Alberto can paddle his canoe a solid 6 MPH in still water. One sunny afternoon he decided to paddle upstream in a river that had a current of $1\frac{1}{2}$ MPH. If he started at 1 PM, how far had he traveled by 2:40 PM? (The current will decrease the speed of the boat going upstream.)

1PM to 2:40 PM is $1\frac{40}{60}$ hr = $1\frac{2}{3}$ hr
downstream: add current to still speed
upstream: subtract current from still speed

$6 - 1\frac{1}{2} = 5\frac{2}{2} - 1\frac{1}{2} = 4\frac{1}{2}$ MPH

$d = rt$
$d = 4\frac{1}{2} \cdot 1\frac{2}{3}$
$= \frac{9}{2} \cdot \frac{5}{3}$
$= \frac{15}{2} = 7\frac{1}{2}$

Alberto would travel $7\frac{1}{2}$ miles upstream

---

2) $8\frac{2}{15}$ **5)** $10\frac{1}{2}$ **8)** $7\frac{1}{2}$ **11)** $1\frac{1}{5}$ **13)** $\frac{15}{16}$ **15)** $4\frac{3}{4}$ pounds

17) 189 pounds **19)** $24\frac{1}{16}$ miles **21)** $2\frac{11}{12}$ pounds

1) $5\frac{3}{4} + 2\frac{7}{12} + 3\frac{1}{18}$

Factor denominators:
- $4 = (2)(2)$
- $12 = (2)(2)(3)$
- $18 = (2)(3)(3)$
- LCD $= (2)(2)(3)(3) = 36$

Note: Each factor is used "the most" that it is used in any one of the terms. $5\frac{3}{4} \cdot \frac{9}{9} + 2\frac{7}{12} \cdot \frac{3}{3} + 3\frac{1}{18} \cdot \frac{2}{2}$

Change to LCD: $5\frac{27}{36} + 2\frac{21}{36} + 3\frac{2}{36}$

$= 10\frac{50}{36}$
$= 10\frac{25}{18}$
$= \boxed{11\frac{7}{18}}$

3) $3\frac{5}{14} + 2\frac{5}{6} + 5\frac{8}{21}$

- $14 = 2 \cdot 7$
- $6 = 2 \cdot 3$
- $21 = 3 \cdot 7$
- LCD $= 2 \cdot 3 \cdot 7 = 42$

$= 3\frac{15}{42} + 2\frac{35}{42} + 5\frac{16}{42}$
$= 10\frac{66}{42}$
$= 10\frac{11}{7} = \boxed{11\frac{4}{7}}$

5) $8\frac{8}{21} - 2\frac{2}{35}$

- $21 = 3 \cdot 7$
- $35 = 5 \cdot 7$
- LCD $= 3 \cdot 5 \cdot 7 = 105$

$= 8\frac{8}{21} \cdot \frac{5}{5} - 2\frac{2}{35} \cdot \frac{3}{3}$
$= 8\frac{40}{105} - 2\frac{6}{105}$
$= \boxed{6\frac{34}{105}}$

7) $8\frac{3}{8} + 5\frac{8}{15} - 7\frac{9}{20}$

- $8 = 2 \cdot 2 \cdot 2$
- $15 = 3 \cdot 5$
- $20 = 2 \cdot 2 \cdot 5$
- LCD $= 2 \cdot 2 \cdot 2 \cdot 3 \cdot 5 = 120$

$= 8\frac{45}{120} + 5\frac{64}{120} - 7\frac{54}{120}$
$= 13\frac{109}{120} - 7\frac{54}{120}$
$= 6\frac{55}{120}$
$= \boxed{6\frac{11}{24}}$

9) $4\frac{3}{14} + 2\frac{9}{35} - 3\frac{3}{20}$

- $14 = 2 \cdot 7$
- $35 = 5 \cdot 7$
- $20 = 2 \cdot 2 \cdot 5$
- LCD $= 2 \cdot 2 \cdot 5 \cdot 7 = 140$

$= 4\frac{30}{140} + 2\frac{36}{140} - 3\frac{21}{140}$
$= 6\frac{66}{140} - 3\frac{21}{140}$
$= 3\frac{45}{140}$
$= \boxed{3\frac{9}{28}}$

11) $8\frac{5}{8} - 5\frac{7}{24} + 2\frac{4}{9}$

- $8 = 2 \cdot 2 \cdot 2$
- $24 = 2 \cdot 2 \cdot 2 \cdot 3$
- $9 = 3 \cdot 3$
- LCD $= 2 \cdot 2 \cdot 2 \cdot 3 \cdot 3 = 72$

$= 8\frac{45}{72} - 5\frac{21}{72} + 2\frac{32}{72}$
$= 3\frac{24}{72} + 2\frac{32}{72}$
$= 5\frac{56}{72}$
$= \boxed{5\frac{7}{9}}$

2) $11\frac{97}{120}$ 4) $12\frac{9}{40}$ 6) $4\frac{101}{120}$ 8) $5\frac{47}{66}$ 10) $2\frac{11}{42}$ 12) $3\frac{17}{40}$

4.8 Answers

1) $5\frac{7}{12} + 2\frac{3}{10} + 3\frac{4}{15}$

$12 = 2 \cdot 2 \cdot 3$
$10 = 2 \cdot 5$
$15 = 3 \cdot 5$
$LCD = 2 \cdot 2 \cdot 3 \cdot 5 = 60$

$= 5\frac{35}{60} + 2\frac{18}{60} + 3\frac{16}{60}$

$= 10\frac{69}{60}$

$= 11\frac{9}{60}$

$= \boxed{11\frac{3}{20}}$

3) $5\frac{3}{20} + 2\frac{7}{12} - 4\frac{5}{18}$

$20 = 2 \cdot 2 \cdot 5$
$12 = 2 \cdot 2 \cdot 3$
$18 = 2 \cdot 3 \cdot 3$
$LCD = 2 \cdot 2 \cdot 3 \cdot 3 \cdot 5 = 180$

$= 5\frac{27}{180} + 2\frac{105}{180} - 4\frac{50}{180}$

$= 7\frac{132}{180} - 4\frac{50}{180}$

$= 3\frac{82}{180}$

$= \boxed{3\frac{41}{90}}$

5) $8\frac{4}{15} - 3\frac{7}{20} + 2\frac{5}{6}$

$15 = 3 \cdot 5$
$20 = 2 \cdot 2 \cdot 5$
$6 = 2 \cdot 3$
$LCD = 2 \cdot 2 \cdot 3 \cdot 5 = 60$

$= 8\frac{16}{60} - 3\frac{21}{60} + 2\frac{50}{60}$
       ←—— combine ——→

$= 10\frac{66}{60} - 3\frac{21}{60}$

$= 7\frac{45}{60}$

$= \boxed{7\frac{3}{4}}$

7) For a special trail mix, Wilbert mixed $2\frac{3}{8}$ pounds of almonds, $3\frac{7}{12}$ pounds of dried apricots and $4\frac{5}{6}$ pounds of dried apples. During his week-long hike, he ate the whole mix. On the average, how much did he eat each day?

$2\frac{3}{8} + 3\frac{7}{12} + 4\frac{5}{6}$

$8 = 2 \cdot 2 \cdot 2$
$12 = 2 \cdot 2 \cdot 3$
$6 = 2 \cdot 3$
$LCD = 2 \cdot 2 \cdot 2 \cdot 3 = 24$

$= 2\frac{9}{24} + 3\frac{14}{24} + 4\frac{20}{24}$

$= 9\frac{43}{24}$

$= 10\frac{19}{24}$ TOTAL weight for 7 days

$10\frac{19}{24} \div 7$

$= \frac{259}{24} \cdot \frac{1}{7}$

$= \boxed{\frac{37}{24}}$ or $\boxed{1\frac{13}{24}}$

Wilbert averaged $1\frac{13}{24}$ pounds per day.

9) Find the result when $6\frac{5}{12}$ is diminished by the product of $1\frac{7}{8}$ and $1\frac{5}{27}$.

$= 6\frac{5}{12} - \left(\frac{15}{8} \cdot \frac{32}{27}\right)$

$= 6\frac{5}{12} - \frac{20}{9}$

$= 6\frac{5}{12} - 2\frac{2}{9}$

$12 = 2 \cdot 2 \cdot 3$
$9 = 3 \cdot 3$
$LCD = 2 \cdot 2 \cdot 3 \cdot 3 = 36$

$= 6\frac{15}{36} - 2\frac{8}{36}$

$= \boxed{4\frac{7}{36}}$

The result is $4\frac{7}{36}$

---

2) $10\frac{29}{120}$ **4)** $3\frac{7}{12}$ **6)** $3\frac{47}{216}$ **8)** $\frac{59}{120}$ pounds   **10)** $36\frac{2}{5}$

1) $2\frac{3}{8} + 4\frac{1}{2} + 3\frac{3}{4}$
$= 2\frac{3}{8} + 4\frac{4}{8} + 3\frac{6}{8}$
$= 9\frac{13}{8}$
$= 9 + 1\frac{5}{8}$
$= \boxed{10\frac{5}{8}}$

4) $3\frac{3}{5} + 1\frac{5}{7} + 3\frac{3}{10}$
$= \frac{18}{5} \cdot \frac{12}{7} + 3\frac{3}{10}$
$= \frac{18}{5} \cdot \frac{7}{12} + 3\frac{3}{10}$
$= \frac{21}{10} + 3\frac{3}{10}$
$= 2\frac{1}{10} + 3\frac{3}{10}$
$= 5\frac{4}{10}$
$= \boxed{5\frac{2}{5}}$

7) $\frac{9}{20} \cdot \left(2\frac{1}{6} + 1\frac{2}{3}\right)$
$= \frac{9}{20} \cdot \left[2\frac{1}{6} + 1\frac{4}{6}\right]$
$= \frac{9}{20} \left[3\frac{5}{6}\right]$
$= \frac{9}{20} \cdot \frac{23}{6}$
$= \boxed{\frac{69}{40}}$ or $\boxed{1\frac{29}{40}}$

10) $7 + 5(x-3) = 2(x+14)$
$7 + 5x - 15 = 2x + 28$
$5x - 8 - 2x = 2x + 28 - 2x$
$3x - 8 + 8 = 28 + 8$
$\frac{3x}{3} = \frac{36}{3}$
$\boxed{x = 12}$

Evaluate each expression using $a = \frac{3}{8}$ $b = $

13) $bn + an$
$\left(\frac{3}{5}\right)\left(\frac{3}{4}\right) + \left(\frac{3}{8}\right)\left(\frac{3}{2}\right)$
$= \frac{3}{5} \cdot \frac{11}{16} + \frac{9}{16} \cdot \frac{5}{5}$
$= \frac{48}{80} + \frac{45}{80}$
$= \boxed{\frac{93}{80}}$ or $\boxed{1\frac{13}{80}}$

15) $ab + x + w$
$= \left(\frac{3}{8}\right)\left(\frac{2}{5}\right) + (1\frac{2}{3}) \div \left(\frac{5}{6}\right)$
$= \frac{3}{20} + \frac{5}{3} \cdot \frac{6}{5}$
$= \frac{3}{20} + 2$
$= \boxed{2\frac{3}{20}}$

17) Find the result when the square of $\frac{2}{3}$ is increased by the product of $1\frac{1}{6}$ and $\frac{4}{21}$.

$= \left(\frac{2}{3}\right)^2 + \left(1\frac{1}{6}\right)\left(\frac{4}{21}\right)$
$= \frac{4}{9} + \frac{7}{6} \cdot \frac{4}{21}$
$= \frac{4}{9} + \frac{2}{9}$
$= \frac{6}{9}$
$= \boxed{\frac{2}{3}}$

The result is $\frac{2}{3}$.

19) Claudia brought a 20 pound watermelon to the Math Picnic. Boris ate $\frac{2}{5}$ of it, Olaf ate $\frac{1}{4}$ of it and Ferd Burfil ate $\frac{1}{10}$ of it. If Tarzan ate $\frac{3}{5}$ of the remainder, how much did he eat?

$\frac{2}{5}(20) = 8$ lbs : amt Boris ate.
$\frac{1}{4}(20) = 5$ lbs : amt Olaf ate.
$\frac{1}{10}(20) = 2$ lbs : " Ferd ate

$8 + 5 + 2 = 15$ lbs total eaten by Boris, Olaf & Ferd.

$20 - 15 = 5$ the remainder

$\frac{3}{5}(5) = 3$

Tarzan ate 3 pounds.

2) $4\frac{7}{8}$ 5) $9\frac{2}{5}$ 8) $17\frac{2}{5}$ 11) $x = 10$ 14) $\frac{1}{6}$ 16) $1\frac{19}{20}$

3) $19\frac{1}{2}$ miles   20) 162 pounds

5.1 Answers

1) $5 + 2(x-3) = 8$
$5 + 2x - 6 = 8$
$2x - 1 + 1 = 8 + 1$
$\dfrac{2x}{2} = \dfrac{9}{2}$

$\boxed{x = \dfrac{9}{2}}$

or

$\boxed{x = 4\tfrac{1}{2}}$

4) $\dfrac{2}{3}x + 12 = 20$
$\dfrac{2}{3}x + 12 - 12 = 20 - 12$
$\dfrac{2}{3}x = 8$
$\dfrac{2}{3}x \cdot \left(\dfrac{3}{2}\right) = 8 \cdot \dfrac{3}{2}$

$\boxed{x = 12}$

7) $\dfrac{7}{8}x - 2 = \dfrac{3}{8}x + 5$
$\dfrac{7}{8}x - 2 - \dfrac{3}{8}x = \dfrac{3}{8}x + 5 - \dfrac{3}{8}x$
$\dfrac{4}{8}x - 2 + 2 = 5 + 2$
$\dfrac{1}{2}x \cdot (2) = 7(2)$

$\boxed{x = 14}$

10) $ax + bx$
$= (2\tfrac{1}{3})(\tfrac{3}{4}) + (1\tfrac{1}{2})(\tfrac{5}{6})$
$= \dfrac{7}{3} \cdot \dfrac{3}{4} + \dfrac{3}{2} \cdot \dfrac{5}{6}$
$= \dfrac{7}{4} + \dfrac{5}{4}$
$= \dfrac{12}{4}$
$= \boxed{3}$

12) $a + b[w - x]$
$= (2\tfrac{1}{3}) + (1\tfrac{1}{2})\left[(\tfrac{2}{3}) - (\tfrac{5}{6})\right]$
$= 2\tfrac{1}{3} + \dfrac{3}{2}\left[\dfrac{4}{6} - \dfrac{5}{6}\right]$
$= 2\tfrac{1}{3} + \dfrac{3}{2}\left[-\dfrac{1}{6}\right]$
$= 2\tfrac{1}{3} - \dfrac{1}{4}$
$= 2\tfrac{4}{12} - \dfrac{3}{12}$
$= \boxed{2\tfrac{1}{12}}$

Solve each of the following word problems:

14) When seven is added to four times a number the result is 25. Find the number.

Let x = the number
$7 + 4x = 25$
$7 + 4x - 7 = 25 - 7$
$\dfrac{4x}{4} = \dfrac{18}{4}$
$\boxed{x = \dfrac{9}{2} \text{ or } 4\tfrac{1}{2}}$

The number is $4\tfrac{1}{2}$.

16) Find two numbers whose sum is 10 and such that five times the smaller increased by twice the larger is 33.

x = smaller number
$10 - x$ = larger "
$5x + 2(10 - x) = 33$
$5x + 20 - 2x = 33$
$3x + 20 - 20 = 33 - 20$
$\dfrac{3x}{3} = \dfrac{13}{3}$
$\boxed{x = \dfrac{13}{3} \text{ or } 4\tfrac{1}{3}}$
$10 - 4\tfrac{1}{3} = \boxed{5\tfrac{2}{3}}$

The numbers are $4\tfrac{1}{3}$ and $5\tfrac{2}{3}$.

18) For the Math BBQ party, Rod Stewart bought 20 pounds of meat. He bought $2\tfrac{1}{2}$ times as much beef as pork, and four more pounds of chicken than beef. How much of each type of meat did he buy?

$\dfrac{5}{2}x$ = amt of beef
x = amt of pork
$\dfrac{5}{2}x + 4$ = " " Chicken

$\dfrac{5}{2}x + x + \dfrac{5}{2}x + 4 = 20$
$6x + 4 - 4 = 20 - 4$
$\dfrac{6x}{6} = \dfrac{16}{6}$
$\boxed{x = \dfrac{8}{3} \text{ or } 2\tfrac{2}{3}}$

$\dfrac{5}{2} \cdot \dfrac{8}{3} = \dfrac{20}{3} \text{ or } \boxed{6\tfrac{2}{3}}$

$\dfrac{5}{2}\left(\dfrac{8}{3}\right) + 4 = \boxed{10\tfrac{2}{3}}$

Rod brought $2\tfrac{2}{3}$ lbs of pork, $6\tfrac{2}{3}$ lbs. of beef and $10\tfrac{2}{3}$ lbs of chicken.

2) $-5\tfrac{1}{3}$ **5)** $x = 48$ **8)** $x = 10$ **11)** 2   **13)** $2\tfrac{1}{2}$ **15)** $2\tfrac{4}{5}$

17) $3\tfrac{9}{10}$ and $4\tfrac{1}{10}$ **19)** $6\tfrac{2}{5}$ miles

5.2 Answers

1) $\frac{2}{3}x + 5 = 7$

$\frac{2}{3}x + 5 - 5 = 7 - 5$

$\frac{2}{3}x \cdot (\frac{3}{2}) = 2(\frac{3}{2})$

$\boxed{x = 3}$

4) $8 + \frac{2}{3}(x-9) = 18$

$8 + \frac{2}{3}x - 6 = 18$

$2 + \frac{2}{3}x - 2 = 18 - 2$

$\frac{2}{3}x \cdot \frac{3}{2} = 16 \cdot \frac{3}{2}$

$\boxed{x = 24}$

7) $3\frac{1}{3} + 2\frac{1}{2}(x-6) = 1\frac{2}{3}$

$3\frac{1}{3} + \frac{5}{2}x - \frac{5}{2} \cdot 6 = \frac{5}{3}$

$3\frac{1}{3} + \frac{5}{2}x - 15 = \frac{5}{3}$

$3\frac{1}{3} + \frac{5}{2}x - 11\frac{2}{3} = \frac{5}{3}$

$-11\frac{2}{3} + \frac{5}{2}x + 11\frac{2}{3} = 1\frac{2}{3} + 11\frac{2}{3}$

$\frac{5}{2}x = 12\frac{4}{3}$

$\frac{5}{2}x \cdot \frac{2}{5} = \frac{40}{3} \cdot \frac{2}{5}$

$\boxed{x = \frac{16}{3}}$ or $\boxed{5\frac{1}{3}}$

---

10. Miriam did some concrete work in her backyard. She used $2\frac{3}{4}$ cubic yards on a walk, $1\frac{1}{2}$ cubic yards on a slab for the hot tub, and $\frac{2}{3}$ cubic yards on the mowing strip. Her concrete company only sold in whole cubic yard amounts. If concrete sells for $37 per cubic yard, what did she pay for the concrete?

$= 2\frac{3}{4} + 1\frac{1}{2} + \frac{2}{3}$

$= 2\frac{9}{12} + 1\frac{6}{12} + \frac{8}{12}$

$= 3\frac{23}{12}$

$= 4\frac{11}{12}$ cu yds → $\boxed{5 \text{ cu yds}}$ only whole #'s

$5(37) = \boxed{185}$

$\boxed{\text{Miriam paid } \$185 \text{ for the concrete}}$

---

12) Find two numbers whose sum is 8, such that five times the smaller diminished by three times the larger is six.

$a = $ smaller number

$8 - a = $ larger "

$5a - 3(8-a) = 6$

$5a - 24 + 3a = 6$

$8a - 24 + 24 = 6 + 24$

$\frac{8a}{8} = \frac{30}{8}$

$a = \frac{15}{4}$

$\boxed{a = 3\frac{3}{4}}$

$7\frac{4}{4} - 3\frac{3}{4} = \boxed{4\frac{1}{4}}$

$\boxed{\text{The smaller number is } 3\frac{3}{4} \text{ and the larger number is } 4\frac{1}{4}}$

---

14) One number is three more than twice another. Five times the smaller added to three times the larger is 15. Find the larger number.

$x = $ smaller number

$2x + 3 = $ larger "

$5x + 3(2x+3) = 15$

$5x + 6x + 9 = 15$

$11x + 9 - 9 = 15 - 9$

$\frac{11x}{11} = \frac{6}{11}$

$\boxed{x = \frac{6}{11}}$

$2(\frac{6}{11}) + 3 = \frac{12}{11} + 3 = 1\frac{1}{11} + 3 = \boxed{4\frac{1}{11}}$

$\boxed{\text{The larger number is } 4\frac{1}{11}}$

---

16) Find two numbers whose sum is three if five times the larger decreased by twice the smaller is 9.

$n = $ smaller number

$3 - n = $ larger "

$5(3-n) - 2n = 9$

$15 - 5n - 2n = 9$

$-7n + 15 - 15 = 9 - 15$

$\frac{-7n}{-7} = \frac{-6}{-7}$

$\boxed{n = \frac{6}{7}}$

$3 - \frac{6}{7} =$

$2\frac{7}{7} - \frac{6}{7} = \boxed{2\frac{1}{7}}$

$\boxed{\text{The numbers are } \frac{6}{7} \text{ and } 2\frac{1}{7}}$

---

2) $4\frac{4}{7}$ **5)** $-2\frac{2}{3}$ **8)** $x = 12$ **11)** $1\frac{1}{9}$ pounds   **13)** $3\frac{1}{2}$ and $5\frac{1}{2}$

15) $10\frac{1}{3}$ pounds   **17)** 122 pounds

5.3 Answers

1) 12 to 18

$$\frac{12}{18} = \boxed{\frac{2}{3}}$$

4) $\frac{2}{3}$ to $1\frac{1}{4}$

$$\frac{\frac{2}{3}}{1\frac{1}{4}} = \frac{2}{3} \div \frac{5}{4}$$
$$= \frac{2}{3} \cdot \frac{4}{5}$$
$$= \boxed{\frac{8}{15}}$$

Alfonso made a fruit salad using the fo strawberries and 6 pounds of apples. U

7) Oranges to bananas

$$\frac{4}{2} = \boxed{\frac{2}{1}}$$

10) Strawberries to the salad

Salad: 4 + 2 + 3 + 6 = 15

$$\frac{3}{15} = \boxed{\frac{1}{5}}$$

Express each of the following as a rate:

13) 105 miles in 3 hours

$$\frac{105 \, mi}{3 \, hr} = \boxed{\frac{35 \, mi}{1 \, hr}}$$

or $\boxed{35 \, MPH}$

Find the unit price of each of the follow

16) $102 for 17 pounds

$$\frac{\$102}{17 \, lb} = \boxed{\frac{\$6}{1 \, lb}}$$

or $\boxed{\$6 \text{ per pound}}$

19) $\frac{2}{3} = \frac{18}{x}$

$$\frac{2x}{2} = \frac{3(18)}{2}^9$$
$$\boxed{x = 27}$$

22) $\frac{7}{8} = \frac{n}{5}$

$$\frac{8n}{8} = \frac{35}{8}$$
$$\boxed{n = \frac{35}{8}} \text{ or } \boxed{4\frac{3}{8}}$$

25) $\frac{\frac{3}{5}}{b} = \frac{\frac{4}{15}}$

$$\frac{3}{4}b = \frac{3}{5} \cdot 15^3$$
$$\frac{4}{3} \cdot \frac{3}{4}b = 9 \cdot \frac{4}{3}$$
$$\boxed{b = 12}$$

28) $\frac{1\frac{3}{4}}{10} = \frac{2\frac{4}{5}}{y}$

$$\frac{7}{4}y = \frac{14}{5} \cdot 10^2$$
$$\frac{4}{7} \cdot \frac{7}{4}y = 28 \cdot \frac{4}{7}$$
$$\boxed{y = 16}$$

31) $\frac{5}{8}x + 7 = 22$

$$\frac{5}{8}x + 7 - 7 = 22 - 7$$
$$\frac{5}{8}x \cdot \frac{8}{5} = 15 \cdot \frac{8}{5}$$
$$\boxed{x = 24}$$

33) $\frac{3}{5}(a+20) = \frac{1}{5}a + 15$

$$\frac{3}{5}a + \frac{3}{5} \cdot 20^4 = \frac{1}{5}a + 15$$
$$\frac{3}{5}a + 12 - \frac{1}{5}a = \frac{1}{5}a + 15 - \frac{1}{5}a$$
$$\frac{2}{5}a + 12 - 12 = 15 - 12$$
$$\frac{2}{5}a \cdot \frac{5}{2} = 3 \cdot \frac{5}{2}$$
$$\boxed{a = \frac{15}{2} \text{ or } 7\frac{1}{2}}$$

35) $\frac{1}{2}(w+8) = \frac{1}{3}(w+18)$

$$\frac{1}{2}w + \frac{1}{2} \cdot 8^4 = \frac{1}{3}w + \frac{1}{3} \cdot 18^6$$
$$\frac{1}{2}w + 4 - \frac{1}{3}w = \frac{1}{3}w + 6 - \frac{1}{3}w$$
$$\frac{3}{6}w - \frac{2}{6}w + 4 - 4 = 6 - 4$$
$$\frac{1}{6}w(6) = 2(6)$$
$$\boxed{w = 12}$$

2) $\frac{4}{5}$ **5)** $\frac{21}{11}$ **8)** $\frac{1}{2}$ **11)** $\frac{2}{5}$ **14)** $\frac{17 \, gallons}{1 \, hour}$ **17)** $18 per kg.

20) $a = 18$ **23)** $4\frac{4}{5}$ **26)** $x = \frac{1}{2}$ **29)** $12\frac{1}{2}$ **32)** $x = 16$ **34)** $n = 16$

36) $x = 200$

5.4 Answers

1) 14 to 35

$$\frac{14}{35} = \boxed{\frac{2}{5}}$$

4) $\frac{3}{4}$ to $\frac{5}{8}$

$$\frac{\frac{3}{4}}{\frac{5}{8}} = \frac{3}{4} \cdot \frac{\cancel{8}^2}{5} = \boxed{\frac{6}{5}}$$

Solve each of the following proportions

7) $\frac{8}{15} = \frac{x}{40}$

$$\frac{15x}{15} = \frac{8 \cdot \cancel{40}^8}{\cancel{15}_3}$$

$$\boxed{x = \frac{64}{3} \text{ or } 21\frac{1}{3}}$$

10) $\frac{7}{12} = \frac{21}{a}$

$$\frac{7a}{7} = \frac{\cancel{12} \cdot \cancel{21}^3}{\cancel{7}}$$

$$\boxed{a = 36}$$

13) $\frac{5}{7} = \frac{x}{3\frac{1}{2}}$

$$7x = 5 \cdot \frac{7}{2}$$

$$7x \cdot \frac{1}{7} = \frac{\cancel{35}^5}{2} \cdot \frac{1}{\cancel{7}}$$

$$\boxed{x = \frac{5}{2} \text{ or } 2\frac{1}{2}}$$

16) The ratio of men to women in a certain school is 5 to 4. If there are 920 men in the school, how many women are there?

let w = number of women

$$\frac{5}{4} = \frac{920}{w} \qquad \frac{\text{men}}{\text{women}} = \frac{\text{men}}{\text{women}}$$

$$\frac{5w}{5} = \frac{4 \cdot \cancel{920}^{184}}{\cancel{5}_1}$$

$$\boxed{w = 736}$$

There are 736 women.

18) The Ferd Burfil Golf Ball Company found that during an average run, there were 9 defective balls out of each 800 balls produced. How many defective balls could be expected out of a production of 25,600 balls?

$\frac{\text{defective}}{\text{production}}$

let d = number of defective balls

$$\frac{9}{800} = \frac{d}{25,600}$$

$$\frac{800d}{800} = \frac{9(25,600)}{800}$$

$$d = 288$$

There were 288 defective balls out of 25,600 balls.

20) Find the unit price for a computer chip if 240 chips cost $4080.

$$\frac{\$4080}{240 \text{ chips}} = \frac{\$17}{1 \text{ chip}}$$

The chips cost $17 each.

22) If a 345 gallon container can be filled in 15 minutes, what is the rate at which the container is filled?

$$\frac{345 \text{ gal}}{15 \text{ min}} = \frac{23 \text{ gal}}{1 \text{ min}}$$

The container is filled at the rate of 23 gallons per minute.

24) $\frac{3}{5}(w - 35) = \frac{1}{5}w + 5$

$$\frac{3}{5}w - \frac{3}{\cancel{5}} \cdot \cancel{35}^7 = \frac{1}{5}w + 5$$

$$\frac{3}{5}w - 21 - \frac{1}{5}w = \frac{1}{5}w + 5 - \frac{1}{5}w$$

$$\frac{2}{5}w - 21 + 21 = 5 + 21$$

$$\frac{\cancel{5}}{\cancel{2}} \cdot \frac{\cancel{2}}{\cancel{5}}w = \cancel{26}^{13} \cdot \frac{5}{\cancel{2}}$$

$$\boxed{w = 65}$$

2) $\frac{5}{1}$ **5)** $\frac{7}{26}$ **8)** $n = 37\frac{1}{2}$ **11)** $3\frac{3}{4}$ **14)** $\frac{3}{4}$ **17)** 32,000 miles

19) $2\frac{9}{20}$ hours or 2 hours 27 minutes   **21)** $41   **23)** 28 computers per hour   **25)** $x = \frac{87}{2}$

5.5 Answers

1) 32 to 48

$$\frac{32}{48} = \boxed{\frac{2}{3}}$$

A serving of a popular cracker contains information to find the following ratios

4) Protein to fat

$$\frac{2}{6} = \boxed{\frac{1}{3}}$$

A coin collection has 8 dimes, 12 nickels ratios:

7) Dimes to nickels

$$\frac{8}{12} = \boxed{\frac{2}{3}}$$

10) Nickels to quarters

$$\frac{12}{20} = \boxed{\frac{3}{5}}$$

Express each of the following as a rate:

13) 336 miles using 16 gallons

$$\frac{336 \text{ miles}}{16 \text{ gal}} = \frac{21 \text{ mi}}{1 \text{ gal}}$$

$$\boxed{21 \text{ miles per gal}}$$

16) 23 pounds for $299

$$\frac{\$299}{23 \text{ pounds}} = \frac{\$13}{1 \text{ lb}}$$

$$\boxed{\$13 \text{ per 1 pound}}$$

Solve the following proportions:

19) $\frac{x}{12} = \frac{42}{72}$

$$\frac{72x}{72} = \frac{\cancel{72}^{7} \cdot \cancel{12}^{1}}{\cancel{72}}$$

$$\boxed{x = 7}$$

22) $\frac{3\frac{1}{2}}{5} = \frac{3}{x}$

$$\frac{7}{2}x = 5 \cdot 3$$

$$\frac{7}{2}x \cdot \frac{2}{7} = 15 \cdot \frac{2}{7}$$

$$\boxed{x = \frac{30}{7} \text{ or } 4\frac{2}{7}}$$

25) $\frac{5}{16}y + 8 = 38$

$$\frac{5}{16}y + 8 - 8 = 38 - 8$$

$$\frac{5}{16}y \left(\frac{16}{5}\right) = 30 \cdot \frac{16}{5}$$

$$\boxed{y = 96}$$

27) $\frac{5}{6}(a-12) = 3$

$$\frac{5}{6}a - \frac{5}{6}(12) = 3$$

$$\frac{5}{6}a - 10 + 10 = 3 + 10$$

$$\frac{5}{6}a \left(\frac{6}{5}\right) = 13\left(\frac{6}{5}\right)$$

$$\boxed{a = \frac{78}{5} \text{ or } 15\frac{3}{5}}$$

29) $\frac{7}{8}(x+16) = \frac{1}{4}(x+32)$

$$\frac{7}{8}x + \frac{7}{8} \cdot 16 = \frac{1}{4}x + \frac{1}{4} \cdot 32$$

$$\frac{7}{8}x + 14 - \frac{1}{4}x = \frac{1}{4}x + 8 - \frac{1}{4}x$$

$$\frac{7}{8}x - \frac{2}{8}x + 14 - 14 = 8 - 14$$

$$\frac{5}{8}x\left(\frac{8}{5}\right) = -6\left(\frac{8}{5}\right)$$

$$\boxed{x = \frac{-48}{5} \text{ or } -9\frac{3}{5}}$$

2) $\frac{7}{1}$ **5)** $\frac{3}{5}$ **8)** $\frac{2}{1}$ **11)** $\frac{1}{7}$ **14)** 14 MPH   **17)** $45 per box

20) $10\frac{1}{2}$ **23)** $\frac{15}{7}$ **26)** $9\frac{3}{5}$ **28)** $x = 48$ **30)** $\frac{8}{5}$ **31)** 279 blue cars

32) 14,325 women   **34)** $2\frac{2}{3}$

1) $5.34 + (16.2)(8.4)$
$5 + (16)(8)$
$5 + 128$
Approximation: $\boxed{133}$

Display: $\boxed{141.42}$

Rounded: $\boxed{141.4}$

4) $15.62 + 4.1 + 3.6$
$16 + 4 + 4$
Approximation: $4 + 4 = \boxed{8}$

Display: $\boxed{7.4097561}$

Rounded: $\boxed{7.4}$

7) $(15.8)(9.725) - (3.2)^2(7.6)$
$(16)(10) - 3^2(8)$
Approximation: $160 - 9(8)$
$160 - 72$
$\boxed{88}$

Display: $\boxed{75.831}$

Rounded: $\boxed{75.8}$

10) Find $\frac{2}{3}$ of 84
$\frac{2}{3} \cdot \overset{28}{84}$
$\boxed{56}$

13) What part of 50 is 24?
let x = the part
$\frac{50x}{50} = \frac{24}{50}$
$x = \boxed{\frac{12}{25}}$

16) $\frac{2}{3}$ of what number is 42?
Let x = the number
$\frac{3}{2} \cdot \frac{2}{3} x = 42 \cdot \frac{3}{2}$
$\boxed{x = 63}$

Solve the following equations:

19) $5 + 3(x + 4) = 20$
$5 + 3x + 12 = 20$
$3x + 17 - 17 = 20 - 17$
$\frac{3x}{3} = \frac{3}{3}$
$\boxed{x = 1}$

22) $2.4x - 4.8 = 31.2$
$2.4x - 4.8 + 4.8 = 31.2 + 4.8$
$\frac{2.4x}{2.4} = \frac{36}{2.4}$
$\boxed{x = 15}$

2) 70.1   5) 80.5   8) 4144.6   11) 24   14) 25   17) $\frac{3}{20}$

20) $x = -8\frac{1}{2}$ 23) $x = 13$

6.2 Answers

1) 45%
   = .45

4) $2\frac{1}{4}$ = 2.25

Write each of the following in common

7) 35%
$\frac{35}{100} = \boxed{\frac{7}{20}}$

10) 37.5% = $\frac{37.5}{100}$
$\frac{375}{1000} = \boxed{\frac{3}{8}}$

Write each of the following in percent f

13) 0.34 = $\boxed{34\%}$

16) $2\frac{1}{2}$ = 2.5 = $\boxed{250\%}$

Estimate the percentage part by shading

19) Shade 23% of
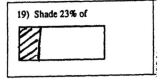

Estimate the percentage "shaded-in" fo

22) 40%
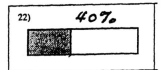

25) 45% of 120 is __54__
$(.45)120 = x$
$\boxed{x = 54}$

28) 16% of __425__ is 68
$\frac{.16x}{.16} = \frac{68}{.16}$
$\boxed{x = 425}$

31) __20__ % of 80 is 16
$\frac{x}{100}(80) = 16$
$\frac{x}{100} \cdot 80 \cdot \frac{100}{80} = 16 \cdot \frac{100}{80}$
$\boxed{x = 20}$

34) 40 is __8__ % of 500
$40 = \frac{x}{100} \cdot 500$
$40 \cdot \frac{100}{500} = \frac{x \cdot 500}{100} \cdot \frac{100}{500}$
$\boxed{x = 8}$

37) 92.6 + (15.3)·(8.6)
$90 + (15)9$
Est: $90 + 135$
$\boxed{225}$

Display: $\boxed{224.18}$

Round: $\boxed{224.2}$

Solve the following word problems:

40) Joe lost 8% of his marbles last Saturday. If he started the day with 175 marbles, how many marbles did he end with?

8% of 175 (marbles) is ___ lost
$(.08)(175) = \boxed{14}$
14 marbles lost
$175 - 14 = \boxed{161}$

Joe ended the day with 161 marbles.

42) To start school, Olaf spent $90 on books, $18 on paper and $12 on a deluxe Mickey Mouse pencil for his math class. What percent of his total did he spend on paper?

$\$90 + \$18 + \$12 = \$120$ TOTAL SPENT

___% of 120 (TOTAL) is 18 (Paper)
$\frac{x}{100} \cdot 120 = 18$
$\frac{x}{100} \cdot 120 \cdot \frac{100}{120} = 18 \cdot \frac{100}{120}$
$\boxed{x = 15}$

Olaf spent 15% of his total on paper.

---

2) 0.085   5) 0.15   8) $\frac{3}{50}$ 11) $\frac{3}{40}$ 14) 3.5%   17) 37.5%

23) About 70%   26) x = 25   29) x = 513   32) x = 48   35) 8.25   38) 1230

41) 48%   43) $189

6.3 Answers

1) 12 is __25__ % of 48

$12 = \frac{x}{100} \cdot 48$

$2\left(\frac{100}{48}\right) = \frac{x}{100} \cdot \frac{48}{1}\left(\frac{100}{48}\right)$

$\boxed{x = 25}$

4) 120% of 65 is __78__

$1.2(65) = x$

$\boxed{x = 78}$

7) 7.5% of 880 is __66__

$(.075)(880) = x$

$\boxed{x = 66}$

10) 40 is __8.3__ % of 480

$(40) = \frac{x}{100} \cdot 480$

$40\left(\frac{100}{480}\right) = \frac{x}{100} \cdot 480 \cdot \left(\frac{100}{480}\right)$

$\boxed{x = 8.333\%}$
$\boxed{x = 8.3}$

---

13) $(16.8)(5.06) + (19.6)(2.6)^2$

$(17)(5) + (20)(3)^2$

Est. $85 + 20(9)$
$85 + 180$
$\boxed{265}$

Display: $\boxed{217.504}$

Round: $\boxed{217.5}$

16) $\frac{(9.62)^2 \cdot \sqrt{104.94}}{55.2}$

Est. $\frac{10^2 \cdot \sqrt{100}}{50}$

$\frac{\overset{2}{100} \cdot 10}{\underset{}{50}} = \boxed{20}$

Display: $\boxed{17.17440090}$

Round: $\boxed{17.2}$

Solve each of the following word problems:

19) Ernie earned $125 per week and got a 6% raise. What is his new salary?

6% of $\underset{Salary}{125}$ is $\underset{raise}{\rule{1cm}{0.15mm}}$

$(.06)(125) = \$7.50$ Raise

$\$125 + \$7.50 = \$132.50$
ORIG WAGE   RAISE   NEW

Ernie has a new salary of $132.50.

---

21) Boris bought a new wastebasket at a 28% discount. If the regular price was $72, what was the price that he paid?

$\underline{28}$ % of $\underset{Reg}{72}$ is $\underset{Discount}{\rule{1cm}{0.15mm}}$

$(.28)(72) = \underset{\underset{Discount}{Amt\,of}}{20.16}$

$\underset{Reg}{\$72} - \underset{Disc.}{\$20.16} = \underset{\underset{Pd.}{Price}}{\$51.84}$

Boris paid $51.84 for the wastebasket.

23) Ferd Burfil was excited to get a 20% discount and pay only $12 for a cap with a propeller on top. What was the regular price of his wonderful cap?

If 20% was discount, then 80% of the price was paid.

80 % of $\rule{0.6cm}{0.15mm}$ is $\underset{Paid}{12}$

$\frac{.80\,x}{.80} = \frac{12}{.80}$

$\boxed{x = 15}$

The regular price was $15.

---

2) 44.1   5) 40   8) 800   11) 40   14) 6.6   17) 4104.6

20) 22.8%   22) $1248   24) $9.62

6.4 Answers

1) 2.6% of 425 is __11.05__

$(.026)425 = x$

$$\boxed{x = 11.05}$$

4) 55% of __2780__ is 1529

$$\frac{.55x}{.55} = \frac{1529}{.55}$$

$$\boxed{x = 2780}$$

7) 18 is __5__ % of 360

$18 = \frac{x}{100} \cdot 360$

$$\frac{18}{3.6} = \frac{3.6x}{3.6}$$

$$\boxed{x = 5}$$

10) 37.5% of 488 is __183__

$(.375)488 = x$

$$\boxed{x = 183}$$

---

13) Olaf sold olives on commission. One week he sold $3680 worth of olives and was paid $294.40 in commission. What was his rate of commission?

$\underline{}$ % of $\underline{\$3680}$ is $\underline{\$294.40}$
              Total sold   Commission

$\frac{x}{100} \cdot 3680 \cdot \frac{100}{3680} = 294.40 \cdot \frac{100}{3680}$

$$x = \boxed{8}$$

He was paid 8% rate of commission.

15) Batman paid $38,750 to have his Batmobile made. He later sold it to a collector who paid $42,500. What was the percent gain on the sale?

$\$42500 - \$38750 = \$3750$
   sold   made   gain

x% of $\underline{38750}$ is $\underline{3750}$
      cost to make   gain

$\frac{x}{100} \cdot 38750 \cdot \frac{100}{38750} = 3750 \cdot \frac{100}{38750}$

$$x = \boxed{9.7}$$

He gained 9.7% on the sale.

17) Ivan scored 95% when he answered 38 questions correctly on a multiple choice history test. If all the questions were of equal value, how many questions were on the test?

$\underline{95\%}$ of $\underline{x}$ is $\underline{38}$
% correct   number on test   number correct

$$\frac{.95x}{.95} = \frac{38}{.95}$$

$$x = \boxed{40}$$

There were 40 questions on the test.

---

2) 250   5) 8600   8) 141.3   11) 2.5%   14) 1200 faucets

16) 237.6 pounds   18) 62%

6.5 Answers

1) 75% of 1024 is __768__

$(.75)1024 = x$

$\boxed{x = 768}$

4) 28% of 430 is __120.4__

$(.28)430 = x$

$\boxed{x = 120.4}$

7) __6.5__% of 560 is 36.4

$\dfrac{x}{100} \cdot 560 = 36.4$

$\dfrac{x}{100} \cdot 560 \cdot \dfrac{100}{560} = 36.4 \cdot \dfrac{100}{560}$

$\boxed{x = 6.5}$

10) $18.6 + (2.3)(879.6)$

Est: $20 + (2)(900)$

$\boxed{1820}$

Display: $\boxed{2041.68}$

Round: $\boxed{2041.7}$

Solve the following word problems

13) A family spent 62% of its budget on essentials, 26% on frivolous items and saved the rest. If their monthly income was $2700, how much money did they save each month?

$\underset{\text{budget}}{100\%} - (\underset{\text{essentials}}{62\%} + \underset{\text{frivolous}}{26\%}) = \underset{\text{save}}{12\%}$

$\underset{\text{saved}}{12\%} \text{ of } \underset{\text{income}}{\$2700} = \underset{\text{money saved}}{x}$

$(.12)2700 = x$

$\boxed{x = \$324}$

The family saved $324 each month

15) Javier sold computer software and was paid 4% for the first $50,000 in sales and 7% for sales over the $50,000 goal. Find his total commission for the month that he sold $74,000 worth of software.

$4\% \text{ of } \underset{\text{1st sales}}{\$50{,}000} = \underset{\text{1st comm.}}{x}$

$.04(50000) = x$

$x = \boxed{\$2000}$ 1st comm.

$\$74{,}000 - \$50{,}000 = \$24{,}000 =$ 2nd sales

$7\% \text{ of } \underset{\text{2nd sales}}{\$24{,}000} = \underset{\text{2nd comm.}}{x}$

$.07(24{,}000) = x$

$x = \boxed{\$1680}$ 2nd comm.

$2000 + 1680 = \boxed{\$3680}$ = total comm.

His total commission is $3680

17) Laurel and Hardy were hired to move 40 pianos. If they moved only 14 pianos in the morning and the rest in the afternoon, what percent of the pianos were moved in the afternoon?

$40 - 14 = 26$ moved in the afternoon

$x\% \text{ of } \underset{\text{total}}{40} \text{ is } \underset{\text{afternoon}}{26}$

$\dfrac{x}{100} \cdot 40 \cdot \dfrac{100}{40} = 26 \cdot \dfrac{100}{40}$

$\boxed{x = 65}$

They moved 65% of the pianos in the afternoon.

2) 37.5%   5) 200   8) 440   11) 65.1   14) $25,824   16) $3,130

18) $68.36   20) $2\tfrac{1}{3}$ and $5\tfrac{2}{3}$ 22) $14.25

7.1 Answers

1) Use the graph below to locate each of the following points:

A (2,4)   B (−3,1)   C (−2,−1)   D (4,0)
E (−2,0) F (3,−2)   G (0,−1)   H (−5,−2)

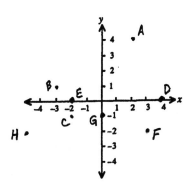

2) Use the graph below to name the coordinate of the points on the graph:

A (−4, 1)   B (4, 0)   C (2, 4)   D (3, −1)
E (0, 2)   F (−3, −1) G (0, 0)   H (−2, −3)

6) The ordinate is four less than one-half the abscissa

$y = \frac{1}{2}x - 4$

7) The ordinate is six more than two-thirds the abscissa

$y = \frac{2}{3}x + 6$

9) The ordinate is three more than five times the abscissa.

(0, 3)   (4, 23)   (−6, −27)   (−3, −12)

$y = 5x + 3$
$y = 5(0) + 3 = 3$
$y = 5(4) + 3 = 23$
$y = 5(-6) + 3 = -27$
$y = 5(-3) + 3 = -12$

12) The ordinate is three less than twice the abscissa.

$y = 2x - 3$

(−2,−7)   (−1,−5)   (0,−3)   (1,−1)   (2, 1)

$y = 2(-2) - 3 = -7$
$y = 2(-1) - 3 = -5$
$y = 2(0) - 3 = -3$
$y = 2(1) - 3 = -1$
$y = 2(2) - 3 = 1$

5) $y = 3x - 2$ 8) $y = 2x^2 + 5$ 10) $y = 3x + 5$ (−2,−1) (−1,2) (0,5) (4,17)

13) a) 17.5 mpg   b) 12 mph   d) 25 mpg

5) The ordinate is two less than three-fourths the abscissa.

(-8, -8)   (-4, -5)   (0, -2)   (4, 1)

$y = \frac{3}{4}x - 2$

$y = \frac{3}{4}(-8) - 2 = -8$

$y = \frac{3}{4}(-4) - 2 = -5$

$y = \frac{3}{4}(0) - 2 = -2$

$y = \frac{3}{4}(4) - 2 = 1$

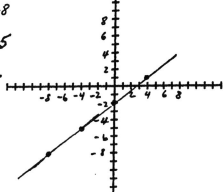

7) The ordinate is the square root of the difference of the abscissa and three.

(3, 0)   (4, 1)   (7, 2)   (12, 3)

$y = \sqrt{x - 3}$

$y = \sqrt{3 - 3} = 0$

$y = \sqrt{4 - 3} = 1$

$y = \sqrt{7 - 3} = 2$

$y = \sqrt{12 - 3} = 3$

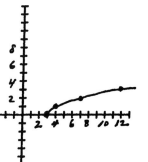

The ordinate is the opposite of the abscissa.

(-3, 3)   (0, 0)   (2, -2)   (5, -5)

$y = -x$

$y = -(-3) = 3$

$y = -(0) = 0$

$y = -(2) = -2$

$y = -(5) = -5$

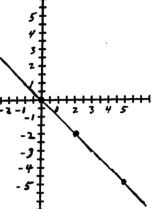

$y = 7x - 4$ 5) $y = 2(x + 1) + 3$ 8) $y = 4x^2 - 6$ 11) $y = x$ 21) 10 AM, $150

7.3 Answers

9) The ordinate is four less than six times the square of the abscissa.

$$f(x) = 6x^2 - 4$$

Translate using function notation, find values of the function using the given values, and graph.

12) Boris is paid a wage of $6 a day plus $5 for each hour worked that day. Let x represent the number of hours worked and write his daily wage as a function of the number of hours worked. Use this function to calculate his wage for the sample number of hours worked.

$$W(x) = 5x + 6$$

$W(2) = 5(2) + 6$ = 16

$W(5) = 5(5) + 6$ = 31

$W(7) = 5(7) + 6$ = 41

$W(8) = 5(8) + 6$ = 46

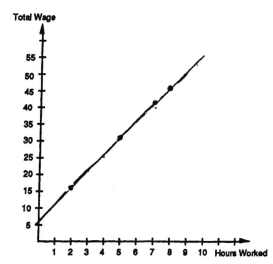

Using the above graph, find each of the following. (DO NOT CALCULATE)

13) Find W(1) = 11

14) Find W(4) = 26

15) Find x if W(x) = 21 $x = 3$ hours

16) If W(x) = 36, how many hours did Boris work?   6 hours

2) $f(x) = 2x - 8$

5) $f(x) = \frac{1}{2}x - 3$

8) $f(x) = \sqrt{x} - 4$

10) $f(x) = 3 \cdot \sqrt{x-2} + 9$

18) $16

21) $88

2) The ordinate is five less than six times the abscissa.

$$g(x) = 6x - 5$$

$g(-4) = 6(-4) - 5 = \boxed{-29}$

$g(3) = 6(3) - 5 = \boxed{13}$

$g(7) = 6(7) - 5 = \boxed{37}$

5) The ordinate is five more than negative two times the abscissa.

$$g(x) = -2x + 5$$

$g(-1) = -2(-1) + 5 = 7$

$g(0) = -2(0) + 5 = 5$

$g(2) = -2(2) + 5 = 1$

$g(4) = -2(4) + 5 = -3$

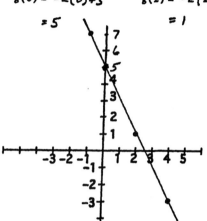

7) The ordinate is three more than twice the abscissa.

$$f(x) = 2x + 3$$

$f(3) = 2(3) + 3 = 9$

$f(-2) = 2(-2) + 3 = -1$

$f(0) = 2(0) + 3 = 3$

$f(1) = 2(1) + 3 = 5$

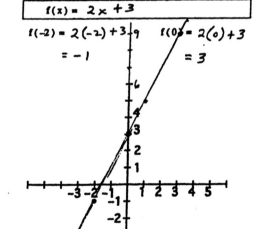

9) For what values of x if any does f(x) = 4

$x = 7$

10) For what values of x if any does f(x) = 0

$x = 2$, $x = -3.5$

11) What is f(4)?

1

$f(x) = 3x + 8$, $f(-5) = -7$, $f(3) = 17$, $f(0) = 8$ 3) $h(x) = \frac{2}{3}x + 1$, $h(-6) = -3$, $h(3) = 3$, $h(6) = 5$

8.1 Answers

Find the area and perimeter of each rec[tangle]

1) l = 12 ft   w = 8 ft

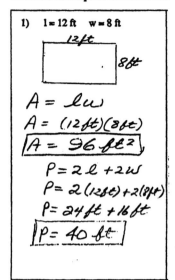

$A = lw$
$A = (12 ft)(8 ft)$
$\boxed{A = 96 ft^2}$

$P = 2l + 2w$
$P = 2(12 ft) + 2(8 ft)$
$P = 24 ft + 16 ft$
$\boxed{P = 40 ft}$

Find the perimeter of each of the follow[ing]

4) Sides are 7 ft, 12 ft and 15 ft.

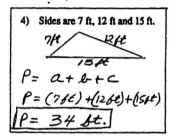

$P = a + b + c$
$P = (7 ft) + (12 ft) + (15 ft)$
$\boxed{P = 34 ft.}$

Find the area of each of the following t[riangles]

7) b = 42 ft, h = 17 ft.

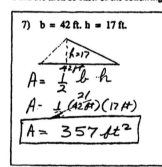

$A = \frac{1}{2} b h$
$A = \frac{1}{2} (42 ft)(17 ft)$
$\boxed{A = 357 ft^2}$

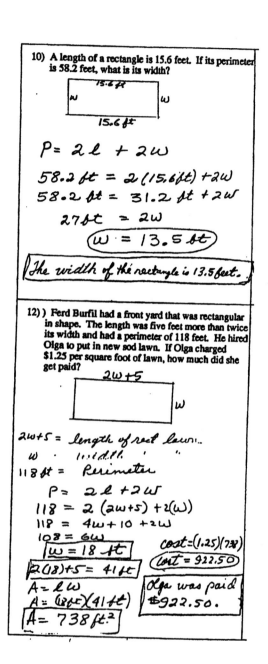

10) A length of a rectangle is 15.6 feet. If its perimeter is 58.2 feet, what is its width?

$P = 2l + 2w$
$58.2 ft = 2(15.6 ft) + 2w$
$58.2 ft = 31.2 ft + 2w$
$27 ft = 2w$
$\boxed{w = 13.5 ft}$

The width of the rectangle is 13.5 feet.

12) Ferd Burfil had a front yard that was rectangular in shape. The length was five feet more than twice its width and had a perimeter of 118 feet. He hired Olga to put in new sod lawn. If Olga charged $1.25 per square foot of lawn, how much did she get paid?

$2w + 5$ = length of rect. lawn
w = width
$118 ft$ = Perimeter

$P = 2l + 2w$
$118 = 2(2w + 5) + 2(w)$
$118 = 4w + 10 + 2w$
$108 = 6w$
$\boxed{w = 18 ft}$
$2(18) + 5 = 41 ft$
$A = lw$
$A = (18 ft)(41 ft)$
$\boxed{A = 738 ft^2}$

cost = (1.25)(738)
$\boxed{cost = 922.50}$

Olga was paid $922.50.

2) A = 364 cm²   P = 80 cm   5) 52 cm   8) A = 636 cm²   11) Area = 193.1 cm²

13) $96.96

8.2 Answers

Find area and circumference of each of

1) r = 5 inches
$A = \pi r^2$
$A = \pi (5)^2$
$A = 25\pi$
$A = 78.5398$
$\boxed{A = 79 \text{ in}^2}$ ← Rounded

$C = 2\pi r$
$C = 2\pi (5)$
$C = 31.4159$
$\boxed{C = 31 \text{ in}}$

Solve each of the following:

4) **Find the area** of a circle whose circumference is 57.2 inches.
$C = 2\pi r$
$57.2 = 2\pi r$
$\frac{57.2}{2\pi} = \frac{2\pi r}{2\pi}$
$\boxed{9.1 = r}$
$A = \pi r^2$
$A = \pi (9.1)^2$
$A = \pi (82.81)$
$A = 260.15528$
$\boxed{A = 260.2 \text{ in}^2}$

7) **Find the base of a triangle** whose height is 24.9 feet and whose area is 450.8 sq. feet.
h = 24.9 ft
$A = \frac{1}{2} b h$
$450.8 = \frac{1}{2} b (24.9)$
$450.8 = 12.45 b$
$\frac{450.8}{12.45} = \frac{12.45 b}{12.45}$
$b = 36.20883$
$\boxed{b = 36.2 \text{ ft}}$

Solve each of the following word problems in the usual man

10) Ivan had a circular flower bed with a radius of 18.2 feet. If he paid Veronica $1.50 per foot to construct a concrete curb around the edge of the flower bed, how much would Ivan pay Veronica?

Find circumference
$C = 2\pi r$
$C = 2\pi (18.2 \text{ ft})$
$C = 114.35397$
$\boxed{C = 114.4 \text{ ft}}$

Cost = ($1.50)(114.4)
= $171.60

Ivan paid Veronica $171.60 for the curb.

12) The Lone Ranger tied his trusty horse Silver to a stake with a rope that was 18 feet long so that the horse could graze in a circular area of Liza's lovely lush lawn while he consumed 3 falafels brimming over with garden fresh tomatoes. **How much grazing area did Silver have?**

r = 18 ft.
$A = \pi r^2$
$A = \pi (18)^2$
$A = \pi (324)$
$A = 1017.876019...$
$\boxed{A = 1018 \text{ ft}^2}$

Silver has a grazing area of 1018 square feet.

*NOTE: Answer is 1017 if π ≈ 3.14 is used.

2) A = 804 m², P = 101 m   5) A = 167.4 m²   8) A = 66.5 cm²   11) 5.3 cm²

13) $120.85 Profit

8.3 Answers

Use a calculator to find the indicated ro

1) $\sqrt{361} = \boxed{19}$

4) $\sqrt{4673} \approx 68.3593..$
$\approx \boxed{68.4}$

7) $\sqrt{96.429} \approx 9.8198..$
$\approx \boxed{9.8}$

For each of the following right triangles into the formula and solve the equation.

10) If a = 5 ft and b = 12 ft find c.

$c^2 = a^2 + b^2$
$c^2 = (5)^2 + (12)^2$
$c^2 = 25 + 144$
$c^2 = 169$
$c = \sqrt{169}$
$\boxed{c = 13 \text{ feet.}}$

13) If a = 12.5 m and c = 32.5 m find b.

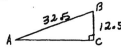

$a^2 + b^2 = c^2$
$(12.5)^2 + b^2 = (32.5)^2$
$156.25 + b^2 = 1056.25$
$b^2 = 900$
$\boxed{b = 30.0 \text{ meters.}}$

Solve the following problems.

16) The base of a ladder is 16.0 feet from the edge of a building. The ladder reaches 38.4 feet up on the building. How tall is the ladder?

$c^2 = a^2 + b^2$
$c^2 = 16^2 + (38.4)^2$
$c^2 = 256 + 1474.56$
$c^2 = 1730.56$
$c = 41.6$

The ladder is 41.6 feet tall.

18) Given rectangle ABCD with DC = 49.2 cm and diagonal AC = 63.4 cm, find the length of the side A

$x^2 + a^2 = c^2$
$x^2 + (49.2)^2 = (63.4)^2$
$x^2 + 2420.64 = 4019.56$
$x^2 = 1599.92$
$x = 39.998999...$
$x \approx 40.0$
$\boxed{AC \approx 40.0 \text{ cm.}}$

20) Romeo has a 24.5 foot ladder. How far from the base of Juliet's house must he place the ladder so that will exactly reach the base of her window, which is 1 feet from the ground?

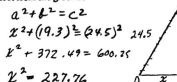

$a^2 + b^2 = c^2$
$x^2 + (19.3)^2 = (24.5)^2$
$x^2 + 372.49 = 600.25$
$x^2 = 227.76$
$x = \sqrt{227.76}$
$x \approx 15.0917$
$\boxed{x \approx 15.1}$

The ladder should be placed 15.1 feet from the base of Juliet's house.

---

2) 56   5) 310.3   8) 31.1   11) c = 7.5 inches   14) a = 17.1 inches

17) 32.2 inches   19) 6.7 miles   21) 28 miles   22) A = 532 ft²

1) Find area and perimeter of the figure below:

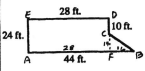

DRAW $CF \perp AB$
$CF = 24 - 10 = 14$
$FB = 44 - 28 = 16$

Area

Plan: $A =$ area rect + area triangle

Formulas: $A = lw + \frac{1}{2}bh$

Substitute: $A = (28)(24) + \frac{1}{2}(16)(14)$

Solve: $A = 672 + 112$

$A = 784$ square feet

Find BC: $(BC)^2 = (14)^2 + (16)^2$
$(BC)^2 = 452$
$BC = \sqrt{452}$
$BC \approx 21.26$
$BC \approx 21$ ft

Perimeter

Plan: $P =$ sum of all sides

Formulas: $P = AB + BC + CD + DE + EA$

Substitute: $P = 44 + 21 + 10 + 28 + 24$

Solve: $P = 127$ feet.

3) Find the area of the figure below:

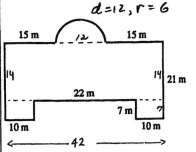

$10 + 22 + 10 = 42$ BOTTOM
$42 - 15 - 15 = 12$ TOP
$21 - 7 = 14$ SIDES

Plan:
$A =$ Area $\frac{1}{2}$ circle + Area large Rect + area 2 small Rect.

$A = \frac{1}{2}\pi r^2 + lw + 2 lw$

$A = \frac{1}{2}\pi(6)^2 + (42)(14) + 2(10)(7)$

$A = \frac{1}{2}\pi(36) + 588 + 140$

$A = 18\pi + 728$

$A \approx 784.54866$

$A = 785 \, m^2$

The area of the figure is 785 square meters.

5) Find the area of the shaded part of the figure. Assume that the circles are tangent to each other and to the sides of the rectangle. (Two figures are tangent if they touch in exactly one point.)

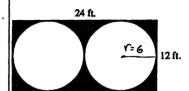

Plan: Area rectangle $-$ 2 Area of circles
$A = lw - 2\pi r^2$
$A = (24)(12) - 2\pi(6)^2$
$A = 288 - 226.1946...$
$A = 288 - 226$
$A = 62$

The area of shaded region is 62 square feet.

7) As part of her dance routine, Isadora placed the foot of a ladder 9.4 feet from the bottom of a wall. If the ladder reached 26.8 feet high on the wall, how long was the ladder?

$C^2 = a^2 + b^2$
$C^2 = (9.4)^2 + (26.8)^2$
$C^2 = 88.36 + 718.24$
$C^2 = 806.6$
$C = \sqrt{806.6}$
$C = 28.4007022$
$C = 28.4$

The ladder was 28.4 feet.

2) $A = 405 \, m^2$, $P = 103 \, m$ 4) $A = 698 \, in^2$, $P = 118$ inches 6) $A = 31 \, in^2$ 8) $s = 23.1 \, cm$

8.5 Answers

1) Convert 28 yards to inches.
$1 \text{ yd} = 3 \text{ ft}$
$1 \text{ ft} = 12 \text{ in}$
$$28 \text{ yds} = 28 \text{ yd} \cdot \frac{3 \text{ ft}}{1 \text{ yd}} \cdot \frac{12 \text{ in}}{1 \text{ ft}}$$
28 yards = 1008 inches.

3) Convert 5 square yards to square feet.
$1 \text{ sq yd} = 9 \text{ sq ft}$
$$5 \text{ yd}^2 = 5 \text{ yd}^2 \cdot \frac{9 \text{ ft}^2}{1 \text{ yd}^2}$$
5 yd² = 45 ft².

5) Convert 7.2 acres to square feet.
$1 \text{ acre} = 43,560 \text{ sq. ft}$
$$7.2 \text{ acres} = 7.2 \text{ acres} \cdot \frac{43,560 \text{ ft}}{1 \text{ acre}}$$
7.2 acres = 313,632 square feet.

7) Convert $2\frac{3}{4}$ miles to yards.
$1 \text{ mile} = 1760 \text{ yards}$
$$2\frac{3}{4} \text{ miles} = 2.75 \text{ mi} \cdot \frac{1760 \text{ yds}}{1 \text{ mi}}$$
$2\frac{3}{4}$ miles = 4840 yards.

9) Convert 15,680 ft² to acres. (Round to nearest hundredths of an acre.)
$1 \text{ acre} = 43,560 \text{ sq. ft}$
$$15,680 \text{ ft}^2 = 15,680 \text{ ft}^2 \cdot \frac{1 \text{ acre}}{43,560 \text{ ft}^2}$$
15,680 ft² = 0.36 acres

11) Convert 185 pounds of fresh water into gallons.
$1 \text{ cu. ft} = 62.5 \text{ pounds}$
$7.5 \text{ gallons} = 1 \text{ cu ft}$
$$185 \text{ pounds} = 185 \text{ lbs} \cdot \frac{1 \text{ ft}^3}{62.5 \text{ lbs}} \cdot \frac{7.5 \text{ gal}}{1 \text{ ft}^3}$$
185 pounds = 22.2 gallons.

13) The Lone Ranger's faithful horse Silver can run 1000 yards in a blistering 2 minutes. What is Silver's speed (to the nearest tenth) in miles per hour?
$1 \text{ mile} = 1760 \text{ yds}$
$1 \text{ hour} = 60 \text{ min}$
$$\frac{1000 \text{ yds}}{2 \text{ min}} = \frac{1000 \text{ yds}}{2 \text{ min}} \cdot \frac{1 \text{ mile}}{1760 \text{ yds}} \cdot \frac{60 \text{ min}}{1 \text{ hr}}$$
$$\frac{1000 \text{ yds}}{2 \text{ min}} = \frac{17.0457... \text{ miles}}{\text{hour}}$$
Silver can run 17.0 MPH.

15) The Batmobile has been clocked at 112 miles per hour on a legal test track. How fast (to the nearest tenth) is that in feet per second?
$1 \text{ mile} = 5280 \text{ ft}$
$1 \text{ hour} = 60 \text{ min}$
$1 \text{ min} = 60 \text{ sec}$
$$112 \text{ MPH} = \frac{112 \text{ mi}}{1 \text{ hr}} \cdot \frac{5280 \text{ ft}}{1 \text{ mi}} \cdot \frac{1 \text{ hr}}{60 \text{ min}} \cdot \frac{1 \text{ min}}{60 \text{ sec}}$$
$112 \text{ MPH} = 164.266...$
The Batmobile goes 164.3 feet per second.

17) Garth Brooks can plow 150 square yards per minute. What is his rate in acres per hour, rounded to the nearest hundredth.
$1 \text{ yd}^2 = 9 \text{ ft}^2$
$1 \text{ acre} = 43,560 \text{ ft}^2$
$1 \text{ hr} = 60 \text{ min}$
$$\frac{150 \text{ yd}^2}{\text{min}} = \frac{150 \text{ yds}^2}{1 \text{ min}} \cdot \frac{9 \text{ ft}^2}{1 \text{ yd}^2} \cdot \frac{1 \text{ acre}}{43,560 \text{ ft}^2} \cdot \frac{60 \text{ min}}{1 \text{ hr}}$$
$$\frac{150 \text{ yd}^2}{\text{min}} = 1.859504..$$
Garth can plow 1.86 acres per hour.

2) 504 hours **4)** 1.7 yards **6)** 31.9 pounds **8)** 408 hours **10)** 0.28 acres **12)** 0.586 ton

14) 270 gallons per hour **16)** 15.7 mph **18)** 23,760 feet per quart

Solve for the indicated value.

1) **Find the area** of a triangle whose base is 12.8 yards and whose height is 6.5 yards.

$A = \frac{1}{2} bh$

$A = \frac{1}{2}(12.8)(6.5)$

$\boxed{A = 41.6 \text{ yd}^2}$

4) The hypotenuse of a right triangle is 31.2 meters and one leg is 12.0 meters. Find the other leg.

$c = 31.2 \text{ m}$

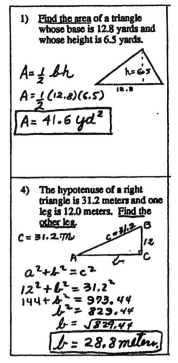

$a^2 + b^2 = c^2$
$12^2 + b^2 = 31.2^2$
$144 + b^2 = 973.44$
$b^2 = 829.44$
$b = \sqrt{829.44}$
$\boxed{b = 28.8 \text{ meters}}$

Solve the following problems in the usual manner.

7) **Find the perimeter** of a right triangle whose legs are 15.2 feet and 21.6 feet.

$c^2 = a^2 + b^2$
$c^2 = (21.6)^2 + (15.2)^2$
$c^2 = 466.56 + 231.04$
$c^2 = 697.6$
$c = \sqrt{697.6}$
$\boxed{c \approx 26.4}$

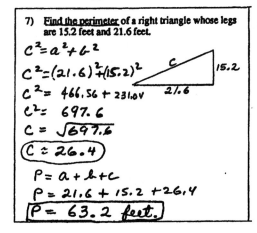

$P = a + b + c$
$P = 21.6 + 15.2 + 26.4$
$\boxed{P = 63.2 \text{ feet}}$

9) The length of a rectangle is one yard more than three times its width. If the perimeter is 122 yards, what is its length?

let x = width
$3x + 1$ = length

$P = 2\ell + 2w$
$122 = 2(3x+1) + 2x$
$122 = 6x + 2 + 2x$
$\frac{120}{8} = \frac{8x}{8}$
$\boxed{x = 15}$ and $\boxed{3(15) + 1 = 46}$
$\boxed{\text{The length is 46 yards.}}$

11) A rectangular shaped garage door measures 16.4 feet long and 8.7 feet high. If one quart of paint covers approximately 90 sq. ft., how many quarts of paint would you need to buy to paint the garage door?

$A = \ell w$
$A = (16.4)(8.7)$
$A = 142.68$
$\boxed{A = 142.7 \text{ ft}^2}$

$\frac{1 \text{ qt}}{90 \text{ ft}^2} = \frac{1 \text{ qt}}{90 \text{ ft}^2} \cdot 142.7 \text{ ft}^2$

$= 1.5685 \text{ qts}$

$\boxed{\text{You would buy 2 quarts of paint.}}$

2) $A = 55.4 \text{ ft}^2$ 5) $h = 10 \text{ ft}^2$ 8) $A = 57.7 \text{ ft}^2$ 10) 15.4 mph

12) Ferd's triangular kite has 1.8 in^2 more area than Fannie's rectangular kite.   13) Cost = $297.36

1) $nab - x$

$\left(\frac{2}{3}\right)(-2)\left(\frac{2}{6}\right) - (-8)$

$-8 + 8$

$\boxed{0}$

4) $nxb + \sqrt{b+a}$

$\left(\frac{2}{3}\right)(-8)\left(\frac{2}{6}\right) + \sqrt{(6)+(-2)}$

$-32 + \sqrt{4}$

$-32 + 2$

$\boxed{-30}$

Solve each of the following equations:

7) $7x - 2 + 4x = 5x + 2$

$11x - 2 - 5x = 5x + 2 - 5x$

$6x - 2 + 2 = 2 + 2$

$\frac{6x}{6} = \frac{4}{6}$

$\boxed{x = \frac{2}{3}}$

$7\left(\frac{2}{3}\right) - 2 + 4\left(\frac{2}{3}\right) = 5\left(\frac{2}{3}\right) + 2$

$\frac{14}{3} - 2 + \frac{8}{3} = \frac{10}{3} + 2$

$\frac{22}{3} - 2 = \frac{16}{3} = \frac{16}{3}$ ✓

10) If $f(x) = 5x - 2$

Find: $f(6) = 5(6) - 2 = \boxed{28}$

$f(-4) = 5(-4) - 2 = \boxed{-22}$

$f\left(\frac{4}{5}\right) = 5\left(\frac{4}{5}\right) - 2 = \boxed{2}$

$f\left(\frac{3}{4}\right) = 5\left(\frac{3}{4}\right) - 2$

$= \frac{15}{4} - \frac{8}{4} = \boxed{\frac{7}{4}}$

$f(0) = 5(0) - 2 = \boxed{-2}$

12) If $h(x) = 4x^2 - 3x + 2$

Find: $h(3) = 4(3)^2 - 3(3) + 2$

$= 36 - 9 + 2$

$\therefore \boxed{h(3) = 29}$

$h(-5) = 4(-5)^2 - 3(-5) + 2$

$= 100 + 15 + 2$

$\boxed{h(-5) = 117}$

$h(6) = 4(6)^2 - 3(6) + 2$

$= 144 - 18 + 2$

$\boxed{h(6) = 128}$

$h(0) = 4(0)^2 - 3(0) + 2$

$= 0 - 0 + 2$

$\boxed{h(0) = 2}$

14) The larger of two numbers is seven more than twice the smaller. Six times the smaller increased by three times the larger is 129. Find the sum of the two numbers.

let x = the smaller
$2x + 7$ = the larger

$6x + 3(2x + 7) = 129$

$6x + 6x + 21 = 129$

$12x + 21 - 21 = 129 - 21$

$\frac{12x}{12} = \frac{108}{12}$

$\boxed{x = 9}$

$2(9) + 7 = 25$

The sum of the numbers is 34.

16) Two-thirds of a number decreased by seven is the same as one-half the number increased by twelve. What is the number?

let x = the number

$\frac{2}{3}x - 7 = \frac{1}{2}x + 12$

$\frac{4}{6}x - 7 - \frac{3}{6}x = \frac{3}{6}x + 12 - \frac{3}{6}x$

$\frac{1}{6}x - 7 + 7 = 12 + 7$

$\frac{1}{6}x \cdot 6 = 19 \cdot 6$

$\boxed{x = 114}$

The number is 114.

18) Add $2\frac{5}{8}$ to the product of $1\frac{1}{6}$ and $3\frac{3}{4}$.

$\left(1\frac{1}{6}\right) \cdot \left(3\frac{3}{4}\right) + 2\frac{5}{8}$

$\frac{7}{6} \cdot \frac{15}{4} + 2\frac{5}{8}$

$\frac{35}{8} + 2\frac{5}{8}$

$4\frac{3}{8} + 2\frac{5}{8}$

$6\frac{8}{8}$

$\boxed{7}$

2) $-11\frac{1}{2}$

5) 9

8) $x = 10$

11) $g(2) = 17$, $g(-3) = -13$, $g\left(\frac{3}{4}\right) = \frac{19}{2}$, $g\left(\frac{-2}{3}\right) = 1$

15) 53 years

17) 275 feet

19) $P(x) = \frac{5}{8}x + 5$

21) 50

23) $x = 80$

9.2 Answers

1) 12% of __65__ is 7.8

$\dfrac{.12x}{.12} = \dfrac{7.8}{.12}$

$\boxed{x = 65}$

4) 3.2% of __245__ is 7.84

$\dfrac{.032x}{.032} = \dfrac{7.84}{.032}$

$\boxed{x = 245}$

Perform the indicated operation:

7) $7\dfrac{3 \cdot 3}{8 \cdot 3} \quad 2\dfrac{5 \cdot 4}{6 \cdot 4} \quad 4\dfrac{3 \cdot 6}{4 \cdot 6}$

$7\dfrac{9}{24} + 2\dfrac{20}{24} - 4\dfrac{18}{24}$

$9\dfrac{29}{24} - 4\dfrac{18}{24}$

$\boxed{5\dfrac{11}{24}}$

Perform the indicated operation:

10) $7\dfrac{5 \cdot 2}{21 \cdot 2} \quad 3\dfrac{3 \cdot 3}{14 \cdot 3} \quad 2\dfrac{1 \cdot 2}{2 \cdot 2}$

$7\dfrac{10}{42} + 3\dfrac{9}{42} - 2\dfrac{21}{42}$

$10\dfrac{19}{42} - 2\dfrac{21}{42}$

$9\dfrac{61}{42} - 2\dfrac{21}{42}$

$7\dfrac{40}{42}$

$\boxed{7\dfrac{20}{21}}$

12) Tarzan rode 13 miles on his elephant, rode 18 miles on his bike and walked 9 miles to visit his 26 year old friend, Boris. What percent of his total distance was on the bike?

total distance = 13 + 18 + 9 = 40 miles

$x\%$ of $\underbrace{40}_{total} = \underbrace{18}_{bike}$

$\dfrac{x}{100} \cdot 40 \cdot \dfrac{100}{40} = 18 \cdot \dfrac{100}{40}$

$\boxed{x = 45}$

45% of his trip was by bicycle

14) Boris was paid $6 per hour plus 8% commission on sales that he made. What was his total earnings the day he worked from 3 PM to 7 PM and sold a total of $280?

8% of $\underbrace{\$280}_{total\ sales} = \underbrace{x}_{commission}$

$.08(280) = x$

$x = \$22.40 =$ commission

4 hours $\cdot$ $6 = \$24 =$ base pay

total earnings
$= \$24 + \$22.40$
$= \$46.40$

Boris earned $46.40

16) Subtract $1\dfrac{4}{5}$ from the product of $6\dfrac{1}{2}$ and $\dfrac{2}{5}$.

$\left(6\dfrac{1}{2}\right)\left(\dfrac{2}{5}\right) - 1\dfrac{4}{5}$

$\left(\dfrac{13}{2}\right)\left(\dfrac{2}{5}\right) - 1\dfrac{4}{5}$

$\dfrac{13}{5} - 1\dfrac{4}{5}$

$2\dfrac{3}{5} - 1\dfrac{4}{5}$

$1\dfrac{8}{5} - 1\dfrac{4}{5} = \boxed{\dfrac{4}{5}}$

The result is $\dfrac{4}{5}$

18) Ivan took a vacation and spent twice as much for his hotel as he did for his airfare. His entertainment and food was $75 more than the cost of the airfare. If the total cost of the trip was $1235, how much did he spend on the hotel?

let x = airfare
$2x$ = hotel
$x + 75$ = entertainment + food

$x + 2x + x + 75 = 1235$
$4x + 75 - 75 = 1235 - 75$
$4x = 1160$
$\boxed{x = 290}$
$2(290) = \boxed{580}$

He spent $580 on the hotel.

20) The width of a rectangle is 12.8 yards. Find the length if the perimeter is 72.4 yards.

let x = length

$2w + 2\ell = p$
$2(12.8) + 2x = 72.4$
$25.6 + 2x - 25.6 = 72.4 - 25.6$
$2x = 46.8$
$\boxed{x = 23.4\ yards}$

The length is 23.4 yards.

---

2) 5000%   **5)** 95   **8)** $1\dfrac{11}{60}$ **11)** $5\dfrac{15}{16}$ **13)** $423.15   **15)** $1\dfrac{3}{4}$ pounds

17) $8\dfrac{2}{3}$ **19)** 19 and 23   **21)** $2\dfrac{1}{3}$ and $3\dfrac{2}{3}$ **22)** $x = 18$

23) $x = -15$ **25)** $x = 24$ **26)** $x = 32$ **28)** $7\dfrac{5}{6}$ **29)** $2\dfrac{3}{8}$

9.3 Answers

1) $4 + 7(x-4) = 6 - 4(6-x)$

$4 + 7x - 28 = 6 - 24 + 4x$

$7x - 24 = 4x - 18$

$7x - 24 - 4x = 4x - 18 - 4x$

$3x - 24 + 24 = -18 + 24$

$3x = 6$

$\boxed{x = 2}$

---

4) 125% of ____ is 70

$\dfrac{1.25x}{1.25} = \dfrac{70}{1.25}$

$\boxed{x = 56}$

---

7) $5\tfrac{2}{3} + 6\tfrac{1}{4}\left(1\tfrac{1}{15}\right)$

$5\tfrac{2}{3} + \dfrac{\cancel{25}^5}{\cancel{4}} \cdot \dfrac{\cancel{16}^4}{\cancel{15}_3}$

$5\tfrac{2}{3} + \dfrac{20}{3}$

$5\tfrac{2}{3} + 6\tfrac{2}{3}$

$11\tfrac{4}{3}$

$\boxed{12\tfrac{1}{3}}$

---

10) $5\tfrac{9}{14} + 2\tfrac{10}{21} - 4\tfrac{5}{6}$

$14 = 2 \cdot 7$

$21 = 3 \cdot 7$

$6 = 2 \cdot 3$

LCD $= 2 \cdot 3 \cdot 7 = 42$

Change to LCD $5\tfrac{27}{42} + 2\tfrac{20}{42} - 4\tfrac{35}{42}$

$7\tfrac{47}{42} - 4\tfrac{35}{42}$

$3\tfrac{12}{42} = \boxed{3\tfrac{2}{7}}$

---

12) Find two numbers whose sum is 25 such that seven times the larger increased by four times the smaller is 151.

let x = smaller
$25 - x$ = larger

$7(25 - x) + 4x = 151$

$175 - 7x + 4x = 151$

$175 - 3x - 175 = 151 - 175$

$-3x = -24$

$\boxed{x = 8}$

$25 - x = 17$

The numbers are 8 and 17

---

14) Find the result when the product of $1\tfrac{3}{4}$ and $2\tfrac{2}{3}$ is subtracted from ten.

$10 - \left(1\tfrac{3}{4}\right)\left(2\tfrac{2}{3}\right)$

$10 - \left(\tfrac{7}{4}\right)\left(\tfrac{8}{3}\right)$

$10 - \tfrac{14}{3}$

$9\tfrac{3}{3} - 4\tfrac{2}{3}$

$\boxed{5\tfrac{1}{3}}$

---

16) Subtract the quotient of $3\tfrac{3}{5}$ and $2\tfrac{1}{4}$ from $8\tfrac{4}{5}$

$8\tfrac{4}{5} - \left[3\tfrac{3}{5} \div 2\tfrac{1}{4}\right]$

$8\tfrac{4}{5} - \left[\tfrac{18}{5} \div \tfrac{9}{4}\right]$

$8\tfrac{4}{5} - \left[\dfrac{\cancel{18}^2}{5} \cdot \dfrac{4}{\cancel{9}}\right]$

$8\tfrac{4}{5} - \tfrac{8}{5}$

$8\tfrac{4}{5} - 1\tfrac{3}{5}$

$\boxed{7\tfrac{1}{5}}$

---

18) Boris bought new size $9\tfrac{1}{2}$ running shoes to help celebrate his arrival at age 21. The shoes had a regular price of $75 but they were on sale at a 25% discount. What was his total cost if the sales tax was 8.5%

25% of $\underset{\text{regular price}}{\$75} = \underset{\text{discount}}{x}$

$.25(75) = x$

$x = \$18.75$ discount

$75 - 18.75 = \$56.25 =$ sales price

$8\tfrac{1}{2}\%$ of $\underset{\text{sales price}}{56.25} = \underset{\text{tax}}{x}$

$.085(56.25) = x$

$x = \boxed{\$4.78 \text{ tax}}$

$56.25 + 4.78 = \boxed{\$61.03}$

Boris paid a total of $61.03 for the shoes

---

2) $x = 5$ **5)** 800%   **8)** $2\tfrac{2}{3}$ **11)** $4\tfrac{37}{120}$ **13)** 7.6 feet   **15)** $334.18

17) $53.72   **19)** $\tfrac{5}{6}$ pounds   **20)** $A = 66.5$ in^2, $c = 28.9$ inches   **22)** $3009.20

Index

A

Abscissa 251

Absolute values 11

Accuracy
    degree of 220

Addition of integers 11

Addition, words for 27

Algebraic expressions 1, 3
    evaluating 15

Approximating answers 217

Approximation symbol 291

Area 281

Area of a circle 287

Area of a rectangle 282

Associative property 45

B

Base numbers 2

Base of triangles 283

C

Calculator 215

Cancellation 120

Categories 83

Chord of a circle 287

Circles 287
    formulas 287

Circumference of a circle 287

Coefficients 2, 28, 43

Common fractions 109

Commutative property 27, 44

Composite integer 113

Constants 2, 43

Conversions, unit 307

Coordinate system 250

Critical digit 219

D

Degree of accuracy 220

Denominators 109
    least common 129
    large 159

Diagonal of a rectangle 296

Diameter of a circle 287

Difference 27

Distributive property 45, 69

Division of integers 14

Division, words for 28

Divisor 113

Drawing graphs 259

E

Equal, words for 28

Exponents 2

Expressions, algebraic
    evaluating 15

Extremes 194

F

Factors 2

Fractions
    addition 122
    division 121
    multiplication 119
    reducing 113, 114
    subtraction 122

Function notation 267

Fundamental Theorem of Fractions 115

G

Graphs 249, 252
    interpreting 275
Greatest common factor 112

H

Height of triangles 283
Higher terms
    building 115
Horizontal direction 250
Hypotenuse 291

I

Identity elements 45
Improper fractions 110, 135
Integers 11
    addition of 11
    division of 14
    multiplication of 13
    subtraction of 12
Inverse properties 46
Irregular figures 299

L

Least common denominators 129
Least common multiple 129
Legs of a right triangle 291
Like terms 44

M

Means 194
Mixed numbers 111, 135
    addition 137
    multiplicaiton and division 136
    subtraction 138
Multiple
    least common 129
Multiples of a number 129
Multiplication
    of integers 13
    words for

N

Negatives 11
Numerators 109
Numerical coefficients 28, 43
Numerical expressions 1

O

Opposites 11
Order of operations 2, 15
Ordinate 251
Origin 250

P

Pentagons 281
Percent 225
Percentage statements 226
Percentage word problems
    solving 227
Perimeter 281
Pi (π) 287
Place values 218

Points 250
Polygons 281
Precision 220
Pricing, unit 193
Prime factors 112, 113
Prime integer 113
Prime tree 114
Product 28
Proper fractions 109
Proportions 194
Pythagorean theorem 291

Q

Quadrilaterals 281
Quotient 28

R

Radius of a circle 287
Rate 192
Ratios 191
    as rate 192
Real numbers 44
Reciprocals 121, 177
    calculator 217
Rectangles 281
    diagonal of 296
Right angles 281
Right triangles 291
Rounding numbers 219

S

Square root
    calculator 216
Square root property 291
Squares 281

Subtraction
    of integers 12
    words for 27
Sum 27
Sum of two numbers 95

T

Tangent figures 305
Terms 43
Terms of proportion 194
Translations 29
Triangles 283
    area formula 283
    perimeter formula 283

U

Unit conversions 307
Unit pricing 193

V

Variables 1, 2, 43
    translating in word problems 29
Vertex of triangles 283
Vertical direction 250

W

Whole numbers 1
Word problems
    5-step process 83
    percent 227
Words for addition 27
Words for divison 28
Words for equal 28
Words for multiplication 28
Words for subtraction 27